PURE AND APPLIED MATHEMATICS

A Program of Monographs, Textbooks, and Lecture Notes

MONOGRAPHS AND TEXTBOOKS IN PURE AND APPLIED MATHEMATICS

1. K. Yano. Integral Formulas in Riemannian Geometry (1970)
2. S. Kobayashi. Hyperbolic Manifolds and Holomorphic Mappings (1970)
3. V. S. Vladimirov. Equations of Mathematical Physics (A. Jeffrey, editor; A. Littlewood, translator) (1970)
4. B. N. Pshenichnyi. Necessary Conditions for an Extremum (L. Neustadt, translation editor; K. Makowski, translator) (1971)
5. L. Narici, E. Beckenstein, and G. Bachman. Functional Analysis and Valuation Theory (1971)
6. D. S. Passman. Infinite Group Rings (1971)
7. L. Dornhoff. Group Representation Theory (in two parts). Part A: Ordinary Representation Theory. Part B: Modular Representation Theory (1971, 1972)
8. W. Boothby and G. L. Weiss (eds.). Symmetric Spaces: Short Courses Presented at Washington University (1972)
9. Y. Matsushima. Differentiable Manifolds (E. T. Kobayashi, translator) (1972)
10. L. E. Ward, Jr. Topology: An Outline for a First Course (1972)
11. A. Babakhanian. Cohomological Methods in Group Theory (1972)
12. R. Gilmer. Multiplicative Ideal Theory (1972)
13. J. Yeh. Stochastic Processes and the Wiener Integral (1973)
14. J. Barros-Neto. Introduction to the Theory of Distributions (1973)
15. R. Larsen. Functional Analysis: An Introduction (1973)
16. K. Yano and S. Ishihara. Tangent and Cotangent Bundles: Differential Geometry (1973)
17. C. Procesi. Rings with Polynomial Identities (1973)
18. R. Hermann. Geometry, Physics, and Systems (1973)
19. N. R. Wallach. Harmonic Analysis on Homogeneous Spaces (1973)
20. J. Dieudonné. Introduction to the Theory of Formal Groups (1973)
21. I. Vaisman. Cohomology and Differential Forms (1973)
22. B.-Y. Chen. Geometry of Submanifolds (1973)
23. M. Marcus. Finite Dimensional Multilinear Algebra (in two parts) (1973)
24. R. Larsen. Banach Algebras: An Introduction (1973)
25. R. O. Kujala and A. L. Vitter (eds). Value Distribution Theory: Part A; Part B. Deficit and Bezout Estimates by Wilhelm Stoll (1973)
26. K. B. Stolarsky. Algebraic Numbers and Diophantine Approximation (1974)
27. A. R. Magid. The Separable Galois Theory of Commutative Rings (1974)
28. B. R. McDonald. Finite Rings with Identity (1974)
29. I. Satake. Linear Algebra (1975)
30. J. S. Golan. Localization of Noncommutative Rings (1975)

Localization of Noncommutative Rings

JONATHAN S. GOLAN

Department of Mathematics
University of Haifa
Haifa, Israel

MARCEL DEKKER, INC. NEW YORK

MARCEL DEKKER, INC.

270 Madison Avenue, New York, New York 10016

LIBRARY OF CONGRESS CATALOG CARD NUMBER: 74-29125

ISBN: 0-8247-6198-7

Current printing (last digit):

10 9 8 7 6 5 4 3 2 1

PRINTED IN THE UNITED STATES OF AMERICA

To my wife Hemda,

קמו בניה ויאשרוה, בעלה ויהללה
רבות בנות עשו חיל, ואת עלית על כלנה

CONTENTS

ACKNOWLEDGMENTS

The author wishes to express his thanks to the University of Florida and to McGill University at which he was visiting assistant professor during the 1971-72 and 1972-73 academic years respectively. During those stays, much of the material in this book was conceived. In particular, he wishes to express his gratitude to Professors Mark Teply of the University of Florida and Joachim Lambek of McGill University for making his stay at their institutions fruitful and challenging and for taking the time to read the first draft of this book and offer incisive comments and helpful suggestions which have been incorporated into the text. The author also wishes to acknowledge the aid of the National Research Council of Canada which partially supported his research during the 1972-73 academic year.

Special thanks are due to the University of Haifa for supporting the preparation of the manuscript of this book, to Mrs. E. Tuval for her dedicated typing of the final draft, and to the editors of Marcel Dekker, Inc. for their patience and understanding.

INTRODUCTION

The purpose of this volume is to present a survey of the work of the past few years on the subject of localization of noncommutative rings with emphasis on results not covered in [90] and [147]. Needless to say, it is impossible to cover all of the ramifications of this rapidly developing field and so the exclusion of various topics on which active research is currently being done was unavoidable.

For the benefit of the reader, all examples have been concentrated in Chapter VII. The reader is urged to refer to that chapter for reification of the concepts introduced in the text. There he will also find many definitions of specific torsion theories some of which are well known and have extensive literatures in their own right while others are essentially unknown and await investigation.

As in most new and developing areas of mathematics, the terminology in the field of localization of noncommutative rings is not yet definitive. The reader is therefore fore-warned that in the literature he is likely to encounter terms vastly different from the ones which I have chosen to use. For example, torsion functors are called "idempotent kernel functors" [63], τ-injective modules are called "divisible" [90], absolutely τ-pure modules are called "closed" [147], etc. Jansian torsion theories are usually refered to in the literature as "TTF-theories".

The reader is assumed to have a knowledge of noncommutative ring theory equivalent to an introductory graduate or advanced undergraduate course, the contents of which correspond roughly to [89]. Homological algebra and category theory are used as little as possible, although a familiarity with the basic terminology ("exact sequence", "functor", etc.) is assumed.

Throughout this volume, the word "ring" is used to refer to an associative ring with unit element 1; R will always denote such a ring. All modules will be assumed to be unitary. If R and S are rings, we will denote the category of left R-modules, the category of right S-modules, and the category of (R,S)-bimodules by R-mod, mod-S, and R-mod-S respectively. If $\mathcal{A}$ is one of these categories, we will often abuse notation and write "$M \in \mathcal{A}$" when we mean to say "M is an object of $\mathcal{A}$". Morphisms in module categories will be written as acting on the side opposite scalar multiplication. All other functions will be written as acting on the left.

If A is a subset of a left R-module M and if N is a submodule of M, we will denote $\{r \in R \mid ra \in N \text{ for every } a \in A\}$

by (N:A). In particular, if $A = \{m\}$ we will write (N:m) for the left annihilator of $m + N \in M/N$.

A submodule N of a left R-module M is small in M if and only if $N + N' = M$ implies that $N' = M$. The sum of all small submodules of M is called the Jacobson radical of M and is denoted by J(M). The Jacobson radical of M is also equal to the intersection of all maximal submodules of M. Dually, a submodule N of M is large in M if and only if $N \cap N' = 0$ implies that $N' = 0$ The intersection of all large submodules of M is called the socle of M and is denoted by soc(M). The socle of M is also the sum of all simple submodules of M.

We will make use of the following set-theoretic conventions: if A is a proper subset of a set B, we will write $A \subset B$; if properness is not assumed or proven, we will write $A \subseteq B$. If $\mathcal{A}$ is a family of subsets of some given set, then $\cup\mathcal{A}$ and $\cap\mathcal{A}$ will respectively denote the union and the intersection of all members of $\mathcal{A}$. If the members of $\mathcal{A}$ are modules, then $\Sigma\mathcal{A}$, $\oplus\mathcal{A}$, and $\Pi\mathcal{A}$ will respectively denote the sum, direct sum, and direct product of the members of $\mathcal{A}$. If M is a module and Ω is a set, then $M^{(\Omega)}$ and M^{Ω} will respectively denote the direct sum and direct product of card(Ω) copies of M. In particular, if card(Ω) $= n < \infty$, we will write M^n instead of $M^{(\Omega)}$. The sets of natural numbers, integers, and rational numbers will be respectively denoted by $\mathbb{N}$, $\mathbb{Z}$, and $\mathbb{Q}$.

CHAPTER I

TORSION THEORIES

1. TORSION AND TORSION-FREE MODULES

A left R-module E is said to be *injective* if and only if the functor $Hom_R(\_,E)$ is exact. It is well known that any left R-module M can be embedded in at least one injective left R-module and, given such a module M, there exists an injective R-module E(M), unique up to isomorphism, which is minimal among all injective left R-modules containing M. The module E(M) is called the *injective hull* of M.

We can partially order the family of all injective left R-modules by setting $E_1 \geq E_2$ if and only if E_1 can be embedded in (and hence is a direct summand of) a direct product of copies of E_2. We say that two injective left R-modules E_1 and E_2 are *equivalent* if and only if $E_1 \geq E_2$ and $E_2 \geq E_1$. This is clearly an equivalence relation.

(1.1) PROPOSITION: *Injective modules* $E_1, E_2 \in$ R-mod

<u>are equivalent if and only if</u>

$\{M \in R\text{-mod} \mid \mathrm{Hom}_R(M,E_1) = 0\} = \{M \in R\text{-mod} \mid \mathrm{Hom}_R(M,E_2) = 0\}$.

PROOF: If E_1 and E_2 are equivalent then the above sets are clearly equal. Conversely, suppose the two sets are equal. There is a canonical R-homomorphism ψ from E_1 into the product of $\mathrm{card}[\mathrm{Hom}_R(E_1,E_2)]$ copies of E_2, defined by $\psi\colon x \mapsto \langle x\alpha \rangle$ where the α range over all $\mathrm{Hom}_R(E_1,E_2)$. If $K = \ker(\psi)$ then $\mathrm{Hom}_R(K,E_2) = 0$ since, by injectivity, any R-homomorphism from K to E_2 can be extended to an R-homomorphism from E_1 to E_2. Therefore, by assumption, $\mathrm{Hom}_R(K,E_1) = 0$. Since K is a submodule of E_1, this means that $K = 0$ and so ψ is an embedding of E_1 into a direct product of copies of E_2. Thus $E_1 \geq E_2$. With reversed subscripts, the same argument shows that $E_2 \geq E_1$ and so we have equivalence. □

An equivalence class of injective left R-modules is called a (<u>hereditary</u>) <u>torsion theory</u> on R-mod. We will denote the collection of all such torsion theories by R-tors.

If $\tau \in$ R-tors we say that a left R-module M is τ-<u>torsion-free</u> if and only if $E(M) \geq E$ for some $E \in \tau$. (It is clear that if this is true for some $E \in \tau$ then it is true for every $E \in \tau$.) That is to say, M is τ-torsion-free if and only if M is embeddable in a member of τ. If

S is the endomorphism ring of E then we have a canonical R-homomorphism $M \to \mathrm{Hom}_S(\mathrm{Hom}_R(M,E),E)$ given by $m \mapsto \lambda_m$ where $\lambda_m(\alpha) = m\alpha$. It is not difficult to check that this is a monomorphism precisely when M is τ-torsion-free. We will denote the class of all τ-torsion-free left R-modules by $\mathcal{F}_\tau$.

If $\tau \in$ R-tors we say that a left R-module M is τ-<u>torsion</u> if and only if $\mathrm{Hom}_R(M,E) = 0$ for some (and hence for every) $E \in \tau$. That is to say, M is τ-torsion if and only if no nonzero homomorphic image of M is τ-torsion-free. We will denote the class of all τ-torsion left R-modules by $\mathcal{T}_\tau$.

(1.2) PROPOSITION: *If* $\tau \in$ R-tors, *then*

(1) $M \in \mathcal{T}_\tau$ *if and only if* $\mathrm{Hom}_R(M,E(N)) = 0$ *for all* $N \in \mathcal{F}_\tau$.

(2) $N \in \mathcal{F}_\tau$ *if and only if* $\mathrm{Hom}_R(M,E(N)) = 0$ *for all* $M \in \mathcal{T}_\tau$.

PROOF: (1) If $M \in \mathcal{T}_\tau$ and $N \in \mathcal{F}_\tau$ then E(N) is embeddable in a direct product of copies of E for some $E \in \tau$. Were $\mathrm{Hom}_R(M,E(N)) \neq 0$ we would then have a nonzero R-homomorphism $M \to E$, a contradiction. Conversely, if $\mathrm{Hom}_R(M,E(N)) = 0$ for all $N \in \mathcal{F}_\tau$ then in particular $\mathrm{Hom}_R(M,E) = 0$ for all $E \in \tau$ and so $M \in \mathcal{T}_\tau$.

(2) If $N \in \mathcal{F}_\tau$ then $\mathrm{Hom}_R(M,E(N)) = 0$ for all $M \in \mathcal{T}_\tau$

by (1). Conversely, assume that $\mathrm{Hom}_R(M,E(N)) = 0$ for all $M \in \mathcal{J}_\tau$. Let $E \in \tau$ and let Ω be a set the cardinality of which equals the cardinality of $\mathrm{Hom}_R(E(N),E)$. Then we again have a canonical R-homomorphism $\psi: E(N) \to E^\Omega$ given by $x \mapsto \langle x\alpha \rangle$ where the α range over all $\mathrm{Hom}_R(E(N),E)$. If $K = \ker(\psi)$ then, as in the proof of Proposition 1.1, we have $\mathrm{Hom}_R(K,E) = 0$ so that $K \in \mathcal{J}_\tau$ and hence, by assumption, $\mathrm{Hom}_R(K,E(N)) = 0$. Since K is a submodule of $E(N)$, this proves that $K = 0$ and thus that $E(N)$ can be embedded in E^Ω, establishing that $N \in \mathcal{F}_\tau$. □

(1.3) PROPOSITION: <u>The correspondences</u> $\tau \mapsto \mathcal{F}_\tau$ <u>and</u> $\tau \mapsto \mathcal{J}_\tau$ <u>are monic</u>.

PROOF: Assume that $\mathcal{F}_\tau = \mathcal{F}_{\tau'}$ and choose $E \in \tau$ and $E' \in \tau'$. Then $E \in \mathcal{F}_\tau = \mathcal{F}_{\tau'}$ so by definition $E \geq E'$. Similarly $E' \geq E$. Therefore E and E' are equivalent and so $\tau = \tau'$.

Now assume that $\mathcal{J}_\tau = \mathcal{J}_{\tau'}$. Then by Proposition 1.2, $\mathcal{F}_\tau = \{N \in R\text{-mod} \mid \mathrm{Hom}_R(M,E(N)) = 0 \text{ for all } M \in \mathcal{J}_\tau\} = \{N \in R\text{-mod} \mid \mathrm{Hom}_R(M,E(N)) = 0 \text{ for all } M \in \mathcal{J}_{\tau'}\} = \mathcal{F}_{\tau'}$ so $\tau = \tau'$. □

We thus see that the classes of modules $\mathcal{F}_\tau$ and $\mathcal{J}_\tau$ each uniquely determines the other and the equivalence class τ. It is therefore natural to ask when a class of left

R-modules is of the form $\mathcal{F}_\tau$ or $\mathcal{T}_\tau$ for some torsion theory τ.

(1.4) PROPOSITION: The following conditions are equivalent for a nonempty class $\mathcal{A}$ of left R-modules:

(1) $\mathcal{A} = \mathcal{F}_\tau$ for some $\tau \in$ R-tors;

(2) $\mathcal{A}$ is closed under taking submodules, injective hulls, direct products, and isomorphic copies.

PROOF: The proof of (1) ⇒ (2) is immediate from the definition of $\mathcal{F}_\tau$. Conversely, assume that $\mathcal{A}$ satisfies (2). Let $\mathcal{A}_0$ be a complete set of representatives of the isomorphism classes of cyclic members of $\mathcal{A}$ and let $E_0 = \Pi\{E(M) \mid M \in \mathcal{A}_0\}$. By (2), $E_0 \in \mathcal{A}$. If $M \in \mathcal{A}$ then by (2) $E(M) \in \mathcal{A}$. For each $x \in E(M)$ we can find a submodule N_x of $E(M)$ such that $E(M) = N_x \oplus E(Rx)$. Clearly $\cap\{N_x \mid x \in E(M)\} = 0$ and so we have a canonical embedding of $E(M)$ into $\Pi\{E(M)/N_x \mid x \in E(M)\} \cong \Pi\{E(Rx) \mid x \in E(M)\}$. This is, in turn, embeddable in a direct product of copies of E_0 and so we have $E(M) \geq E_0$ for all $M \in \mathcal{A}$.

Conversely, if $E(M) \geq E_0$ then M is embeddable in a direct product of copies of E_0 and so, by (2), we must have $M \in \mathcal{A}$. Therefore $\mathcal{A} = \{M \in \text{R-mod} \mid E(M) \geq E_0\} = \mathcal{F}_\tau$ where τ is the equivalence class of E_0. □

For any class $\mathcal{A}$ of left R-modules, we can consider the complete set $\mathcal{A}_0$ of representatives of isomorphism classes of cyclic submodules of members of $\mathcal{A}$ and form the injective left R-module $E_0 = \Pi\{E(M) \mid M \in \mathcal{A}_0\}$. We denote the equivalence class of E_0 by $\chi(\mathcal{A})$ and call it the torsion theory <u>cogenerated</u> by $\mathcal{A}$. We always have $\mathcal{A} \subseteq \mathcal{F}_{\chi(\mathcal{A})}$. If $\mathcal{A} = \{M\}$ we write $\chi(M)$ instead of $\chi(\mathcal{A})$ and note that $\chi(M)$ is precisely the equivalence class of $E(M)$.

A class $\mathcal{A}$ of left R-modules is said to be <u>closed under extensions</u> if and only if, for every exact sequence $0 \to M' \to M \to M'' \to 0$ in R-mod, $M', M'' \in \mathcal{A}$ implies $M \in \mathcal{A}$. As a direct consequence of Proposition 1.2 we then have the following result.

(1.5) PROPOSITION: <u>If</u> $\tau \in$ R-tors <u>then</u> $\mathcal{F}_\tau$ <u>is closed under extensions</u>.

PROOF: Let $0 \to N' \to N \to N'' \to 0$ be an exact sequence in R-mod with $N', N'' \in \mathcal{F}_\tau$. Let $M \in \mathcal{T}_\tau$ and let $0 \neq \alpha \in \mathrm{Hom}_R(M,E(N))$. Then $M\alpha \cap N \neq 0$ since N is large in $E(N)$, and so there exists an $m \in M$ with $0 \neq Rm\alpha \in N$. From the definition of $\mathcal{T}_\tau$ it is immediate that $Rm \in \mathcal{T}_\tau$ and so, by Proposition 1.2, $Rm\alpha \cap N' = 0$ since $N' \in \mathcal{F}_\tau$. Therefore α induces a nonzero homomorphism $Rm \to N/N' \cong N''$, which is a contradiction since $N'' \in \mathcal{F}_\tau$. We deduce, therefore, that we must have $\mathrm{Hom}_R(M,E(N)) = 0$ for every

$M \in \mathcal{J}_\tau$ proving, by Proposition 1.2, that $M \in \mathcal{F}_\tau$. □

(1.6) PROPOSITION: The following conditions are equivalent for a nonempty class $\mathcal{A}$ of left R-modules:

(1) $\mathcal{A} = \mathcal{J}_\tau$ for some $\tau \in$ R-tors;

(2) $\mathcal{A}$ is closed under taking submodules, homomorphic images, direct sums, and extensions.

PROOF: The proof of (1) ⇒ (2) is immediate from the definition of $\mathcal{J}_\tau$ and the exactness of the functor $\mathrm{Hom}_R(\_,E)$ for any injective left R-module E. Conversely, assume that $\mathcal{A}$ satisfies (2). Let $\mathcal{B} = \{N \in \text{R-mod} \mid \mathrm{Hom}_R(M,E(N)) = 0$ for all $M \in \mathcal{A}\}$ and let $E_0 \in \chi(\mathcal{B})$. Then $\mathrm{Hom}_R(M,E_0) = 0$ for every $M \in \mathcal{A}$ and so $\mathcal{A} \subseteq \mathcal{J}_{\chi(\mathcal{B})}$. Conversely, let $N \in \mathcal{J}_{\chi(\mathcal{B})}$ and let $\{x_i \mid i \in \Omega\}$ be the set of all elements x of N with $Rx \in \mathcal{A}$. Then $\Omega \neq \emptyset$ since $R0 = 0 \in \mathcal{A}$ by (2). Moreover, $N' = \Sigma Rx_i$ is a homomorphic image of $\oplus Rx_i$ and so belongs to $\mathcal{A}$ by (2). Furthermore, if $x \in N \setminus N'$ and if $[Rx + N']/N' \in \mathcal{A}$, then by the exactness of the sequence $0 \to N' \to Rx + N' \to [Rx + N']/N' \to 0$ it follows that $Rx + N' \in \mathcal{A}$, contradicting the definition of N'. Thus we conclude that no nonzero submodule of N/N' belongs to $\mathcal{A}$.

Since N/N' is a large submodule of E(N/N') it

follows that no nonzero submodule of $E(N/N')$ belongs to $\mathcal{A}$. Since $\mathcal{A}$ is closed under taking homomorphic images, this implies in particular that $\mathrm{Hom}_R(M, E(N/N')) = 0$ for all $M \in \mathcal{A}$. By Proposition 1.2, $E(N/N') \in \mathcal{F}_{\chi(\mathcal{B})}$. Therefore $\mathrm{Hom}_R(N, E(N/N')) = 0$, which is a contradiction unless $N = N'$. Therefore we have shown that $N \in \mathcal{A}$, proving that $\mathcal{A} = \mathcal{T}_{\chi(\mathcal{B})}$. □

If $\mathcal{A}$ is any class of left R-modules we denote the torsion theory cogenerated by $\{N \in R\text{-mod} \mid \mathrm{Hom}_R(M, E(N)) = 0$ for all $M \in \mathcal{A}\}$ by $\xi(\mathcal{A})$ and call it the torsion theory <u>generated</u> by $\mathcal{A}$. We always have $\mathcal{A} \subseteq \mathcal{T}_{\xi(\mathcal{A})}$. If $\mathcal{A} = \{M\}$ we write $\xi(M)$ instead of $\xi(\mathcal{A})$.

(1.7) PROPOSITION: *If* $\tau \in R\text{-tors}$, $M \in \mathcal{F}_\tau$, *and* $M/N \in \mathcal{T}_\tau$, *then* N *is a large submodule of* M.

PROOF: If $0 \neq m \in M$, then $[Rm + N]/N \cong Rm/[Rm \cap N]$, and this is τ-torsion since M/N is τ-torsion. Since M is τ-torsion-free, Rm is τ-torsion-free and so we must have $Rm \cap N \neq 0$. Therefore, N is large in M. □

<u>References for Section 1</u>

Dickson [36]; Faith [45]; Findlay and Lambek [47, 48]; Gabriel [52]; Goldman [63]; Jans [79]; Lambek [90, 92, 93]; Mishina and Skornyakov [104]; Popescu [119]; Stenström [147]; Walker and Walker [164].

2. THE TORSION FUNCTOR

From the proof of Proposition 1.6 we can isolate another important property of torsion theories.

(2.1) PROPOSITION: *Let* $\tau \in$ R-tors *and let* $M \in$ R-mod. *Then there exists a unique submodule* M' *of* M *satisfying*:

(1) $M' \in \mathcal{T}_\tau$;

(2) $M/M' \in \mathcal{F}_\tau$.

PROOF: Take $M' = \Sigma\{Rm \mid m \in M \text{ and } Rm \in \mathcal{T}_\tau\}$. □

We denote the submodule M' of a left R-module M described in Proposition 2.1 by $T_\tau(M)$ and call it the τ-*torsion submodule* of M.

(2.2) PROPOSITION: *If* $\tau \in$ R-tors *and* $M \in$ R-mod, *then*

(1) $T_\tau(M) = M$ *if and only if* M is τ-*torsion*;

(2) $T_\tau(M) = 0$ *if and only if* M is τ-*torsion-free*;

(3) *If* $\alpha \in \mathrm{Hom}_R(M,N)$ *then* $T_\tau(M)\alpha \subseteq T_\tau(N)$;

(4) $T_\tau(M/T_\tau(M)) = 0$;

(5) *If* N *is a submodule of* M *then* $T_\tau(M) \cap N = T_\tau(N)$.

PROOF: The conclusions follow directly from the definitions. □

In particular, $T_\tau(\_)$ is a left exact subfunctor of the identity functor on R-mod which we will call the τ-torsion-functor.

(2.3) PROPOSITION: The correspondence $\tau \mapsto T_\tau$ is monic.

PROOF: If $T_\tau = T_{\tau'}$, then $\mathcal{J}_\tau = \{M \in \text{R-mod} \mid T_\tau(M) = M\} = \{M \in \text{R-mod} \mid T_{\tau'}(M) = M\} = \mathcal{J}_{\tau'}$, and so $\tau = \tau'$ by Proposition 1.3. □

Thus the functor T_τ uniquely determines τ and so we want to ask when a subfunctor of the identity functor on R-mod is T_τ for some $\tau \in$ R-tors.

(2.4) PROPOSITION: The following conditions are equivalent for a submodule N of a left R-module M:

(1) There exists a $\tau \in$ R-tors with $T_\tau(M) = N$.

(2) For every $x \in N$ and $y \in M \setminus N$, $(0:x) \not\subseteq (N:y)$.

PROOF: (1) ⇒ (2): Suppose that there exists an $x \in N$ and $y \in M \setminus N$ with $(0:x) \subseteq (N:y)$. Then we can define a nonzero R-homomorphism $\alpha: Rx \to M/N$ by $ax \mapsto ay + N$. But if (1) holds then $Rx \in \mathcal{J}_\tau$ and $M/N \in \mathcal{F}_\tau$, so there can be no nonzero R-homomorphism between them. Thus we have (2).

(2) ⇒ (1): Let $\tau = \xi(N)$. Then surely $N \subseteq T_\tau(M)$. If we do not have equality, there exists a submodule N' of N and a nonzero R-homomorphism $\alpha: N' \to T_\tau(M)/N$. But then there exist an $x \in N' \subseteq N$ and a $y \in T_\tau(M) \setminus N$ such that $0 \neq x\alpha = y + N$, whence $(0:x) \subseteq (N:y)$, which is a contradiction. □

(2.5) PROPOSITION: *The following conditions are equivalent for a subfunctor* F *of the identity functor on* R-mod:

(1) $F = T_\tau$ *for some* $\tau \in$ R-tors.

(2) F *is left exact and* $F(M/F(M)) = 0$ *for all* $M \in$ R-mod.

PROOF: The proof of (1) ⇒ (2) follows directly from Proposition 2.2. Conversely, if F is a subfunctor of the identity functor on R-mod which satisfies (2), let $\mathcal{A} = \{M \in \text{R-mod} \mid F(M) = M\}$. Then $\mathcal{A}$ is clearly closed under taking submodules, direct sums, and isomorphic copies. If $M \in \mathcal{A}$ and N is a submodule of M, then the canonical homomorphism $M \to M/N$ induces a homomorphism $F(M) \to F(M/N)$ with kernel $N \cap F(M) = N$. This yields a monomorphism $F(M)/N \to F(M/N)$ and so $M/N \in \mathcal{A}$. Thus $\mathcal{A}$ is closed under taking homomorphic images.

Finally, if N is a submodule of M and if N and M/N are both members of $\mathcal{A}$, then $F(M) \cap N = F(N) = N$ and so N is a submodule of F(M). Therefore M/F(M) is a homomorphic image of M/N and so belongs to $\mathcal{A}$. Thus $M/F(M) = F(M/F(M)) = 0$ and so $M = F(M)$. Thus $M \in \mathcal{A}$.

By Proposition 1.6, there exists a $\tau \in$ R-tors with $\mathcal{A} = \mathcal{J}_\tau$ and so $F = T_\tau$. □

By Proposition 2.5 we see that in order for a subfunctor F of the identity functor on R-mod to be of the form T_τ for some $\tau \in$ R-tors, left exactness alone does not suffice. However, from left exact subfunctors we can build subfunctors having this property.

(2.6) PROPOSITION: <u>Let</u> F <u>be a left exact subfunctor of the identity functor on</u> R-mod <u>and let</u> $\mathcal{A} = \{M \in \text{R-mod} \mid F(M) = M\}$. <u>Let</u> $\overline{F}$ <u>be the subfunctor of the identity functor on</u> R-mod <u>defined by</u> $\overline{F}(M) = \cap\{{}_RN \subseteq M \mid F(M/N) = 0\}$. <u>Then</u>

(1) $F(M) \subseteq \overline{F}(M)$ <u>for every left</u> R-<u>module</u> M;

(2) <u>If</u> $F(M) = 0$ <u>then</u> $\overline{F}(M) = 0$ <u>for every left</u> R-<u>module</u> M;

(3) $F(M/\overline{F}(M)) = 0$ <u>for every left</u> R-<u>module</u> M;

(4) $\overline{\mathcal{A}} = \{M \in \text{R-mod} \mid \overline{F}(M) = M\}$ <u>is closed under taking submodules and homomorphic images</u>;

(5) $\overline{F} = T_{\xi(\mathcal{A})}$.

PROOF: (1) If $F(M/N) = 0$ then $F(M)$ is contained in the kernel of the canonical R-epimorphism

$M \to M/N$ whence $F(M) \subseteq \overline{F}(M)$.

(2) If $F(M) = F(M/0) = 0$, then by definition of $\overline{F}$ we have $\overline{F}(M) \subseteq 0$ and so $\overline{F}(M) = 0$.

(3) Suppose that N is a submodule of M with $F(M/N) = 0$. Then $\overline{F}(M) \subseteq N$ and $N' = N/\overline{F}(M)$ is a submodule of $M' = M/\overline{F}(M)$ with $F(M'/N') = 0$. Thus $\overline{F}(M') \subseteq N'$ for any such submodule N' of M'. However, by definition of $\overline{F}$, we have $\cap N' = 0$ and so $F(M') = F(M/\overline{F}(M)) = 0$.

(4) Assume that $\overline{F}(M) = M$ and let N be a submodule of M. Then $\overline{F}(M/N) \subseteq M/N = \overline{F}(M)/N$. Conversely, let $N' = \{m \in M \mid m + N \in \overline{F}(M/N)\}$. Then the canonical surjection $M \to M/N$ induces a monomorphism $M/N' \to (M/N)/\overline{F}(M/N)$. But $F[(M/N)/\overline{F}(M/N)] = 0$ by (3), so by left exactness $F(M/N') = 0$ whence $\overline{F}(M) \subseteq N'$. Therefore $M/N = \overline{F}(M)/N \subseteq \overline{F}(M/N)$, so we have $\overline{F}(M/N) = M/N$. This proves the closure of $\overline{\mathcal{A}}$ under taking homomorphic images.

Now let $N' = N/\overline{F}(N)$ and let $M' = M/\overline{F}(N)$. Then $F(N') = 0$ by (3) and $F(M') = M'$ since M' is a homomorphic image of M. Thus $M' \in \mathcal{A}$ and so by left exactness $N' \in \mathcal{A}$ whence $F(N') = N'$. Therefore $N' = 0$, so $N = \overline{F}(N)$, proving that $\overline{\mathcal{A}}$ is closed under taking submodules.

(5) We first claim that $\overline{F} = T_\tau$ for some $\tau \in$ R-tors. By (2) and (3), $\overline{F}(M/\overline{F}(M)) = 0$ for every $M \in$ R-mod, and so

by Proposition 2.5 it suffices to show that $\bar{F}$ is left exact. Let N be a submodule of a left R-module M. Then we have a canonical monomorphism $N/[N \cap \bar{F}(M)] \to M/\bar{F}(M)$. Since $F(M/\bar{F}(M)) = 0$ we have $F(N/[N \cap \bar{F}(M)]) = 0$ by the left exactness of F and so $\bar{F}(N) \subseteq N \cap \bar{F}(M)$. By (4), $N \cap \bar{F}(M) \in \bar{\mathcal{A}}$ and so $N \cap \bar{F}(M) = \bar{F}(N \cap \bar{F}(M)) \subseteq \bar{F}(N)$ whence $\bar{F}(N) = N \cap \bar{F}(M)$. This shows that $\bar{F}$ is left exact and hence that $\bar{F} = T_\tau$ for some $\tau \in$ R-tors.

That $\tau = \xi(\mathcal{A})$ follows from Proposition 1.3 and the definition of $\xi(\mathcal{A})$. □

Given a left exact subfunctor F of the identity functor on R-mod, we can also define $\bar{F} = T_{\xi(\mathcal{A})}$ using an Amitsur transfinite construction. For each ordinal i, define the subfunctor F_i of the identity functor on R-mod as follows:

(1) $F_0(M) = 0$ for all $M \in$ R-mod.

(2) If i is not a limit ordinal, $F_i(M)/F_{i-1}(M) = F(M/F_{i-1}(M))$ for all $M \in$ R-mod.

(3) If i is a limit ordinal, $F_i(M) = \cup\{F_j(M) \mid j < i\}$ for all $M \in$ R-mod.

Then for each left R-module M there exists a smallest ordinal $k(M)$, which is not a limit ordinal, satisfying the condition that $F_{k(M)}(M) = F_i(M)$ for all ordinals $i \geq k(M)$. We call $k(M)$ the F-<u>length</u> of M.

(2.7) PROPOSITION: Let F be a left exact subfunctor of the identity functor on R-mod and let $\mathcal{A} = \{M \in \text{R-mod} \mid F(M) = M\}$. Let $\overline{F}$ and $\langle F_i \rangle$ be the associated functors defined above. Then for any left R-module M, $\overline{F}(M) = F_{k(M)}(M)$.

PROOF: By Proposition 2.6 it suffices to show that $M \in \mathcal{T}_{\xi(\mathcal{A})}$ if and only if $M = F_{k(M)}(M)$. Assume that $M \in \mathcal{T}_{\xi(\mathcal{A})}$ and that $M \neq F_{k(M)}(M)$. Then $0 \neq M/F_{k(M)}(M) \in \mathcal{T}_{\xi(M)}$, so $F_{k(m)+1}(M)/F_{k(M)} = F(M/F_{k(M)}(M)) \neq 0$ whence $F_{k(M)+1}(M) \neq F_{k(M)}(M)$, which is a contradiction.

Conversely, assume that $M = F_{k(M)}(M)$ and $M \notin \mathcal{T}_{\xi(\mathcal{A})}$. Then there exists an ordinal i minimal with respect to $F_i(M) \not\subseteq T_{\xi(\mathcal{A})}(M)$. This i is clearly not a limit ordinal. Therefore, $F_{i-1}(M) \subseteq T_{\xi(\mathcal{A})}(M)$, and so there exists a nonzero R-homomorphism $\alpha: F_i(M)/F_{i-1}(M) \to M/T_{\xi(\mathcal{A})}(M)$. But $F_i(M)/F_{i-1}(M) = F(M/F_{i-1}(M)) \in \mathcal{A}$, and so $\text{im}(\alpha) \in \mathcal{A}$. Thus $M/T_{\xi(\mathcal{A})}(M)$ has a nonzero submodule belonging to $\mathcal{A}$, which is a contradiction since $\mathcal{A} \subseteq \mathcal{T}_{\xi(\mathcal{A})}$. Thus $M \in \mathcal{T}_{\xi(\mathcal{A})}$. □

We now mention a very important example of a functor of the type discussed in Propositions 2.6 and 2.7. For any ring R, let R-simp be a complete set of representatives of the isomorphism classes of simple left R-modules. If $\mathcal{A} \subseteq$ R-simp we can define a subfunctor F of the identity functor on

R-mod by $F(M) = \oplus \{{}_R N \subseteq M \mid N$ is isomorphic to a member of $\mathcal{A}\}$. Then F is clearly left exact. We therefore obtain a transfinite sequence $\langle F_i \rangle$ called the $\mathcal{A}$-Loewy sequence on R-mod. The F-length of a left R-module M is called the $\mathcal{A}$-Loewy length of M. If $\mathcal{A}$ = R-simp we speak simply of the Loewy sequence and Loewy length.

References for Section 2

Dickson [36]; Fuchs [50]; Gabriel [52]; Goldman [63]; Lambek [90]; Maranda [98]; Mishina and Skornyakov [104]; Shores [139, 141]; Stenström [147]; Walker and Walker [164].

3. DENSE AND PURE SUBMODULES

If $\tau \in$ R-tors then a submodule N of a left R-module M is said to be τ-dense in M if and only if M/N is τ-torsion. In particular, we will be interested in the set of all τ-dense left ideals of the ring R, which we will denote by $\mathcal{L}_\tau$.

(3.1) PROPOSITION: *The correspondence $\tau \mapsto \mathcal{L}_\tau$ is monic.*

PROOF: Assume that $\mathcal{L}_\tau = \mathcal{L}_{\tau'}$. Then $\mathcal{T}_\tau = \{M \in \text{R-mod} \mid (0:m) \in \mathcal{L}_\tau$ for every $m \in M\} = \{M \in \text{R-mod} \mid (0:m) \in \mathcal{L}_{\tau'}$ for every $m \in M\} = \mathcal{T}_{\tau'}$, and so $\tau = \tau'$ by Proposition 1.3. □

We therefore see that the sets $\mathcal{L}_\tau$ uniquely determine their respective torsion theories τ and so would like to find out when a set of left ideals of R is of the form $\mathcal{L}_\tau$ for some $\tau \in$ R-tors.

A nonempty set $\mathcal{A}$ of left ideals of R is called an <u>idempotent filter</u> provided the following conditions are satisfied:

(1) If $I \in \mathcal{A}$ and $a \in R$ then $(I:a) \in \mathcal{A}$;

(2) If I is a left ideal of R for which there exists an $H \in \mathcal{A}$ such that $(I:a) \in \mathcal{A}$ for all $a \in H$ then $I \in \mathcal{A}$.

(3.2) PROPOSITION: *If $\mathcal{A}$ is an idempotent filter of left ideals of R then the following conditions are satisfied:*

(3) *If $I \in \mathcal{A}$ and H is a left ideal of R containing I then $H \in \mathcal{A}$.*

(4) *If $I, H \in \mathcal{A}$ then $I \cap H \in \mathcal{A}$.*

PROOF: If $I \in \mathcal{A}$ and H is a left ideal of R containing I, then for each $a \in I$, we have $(H:a) = R$. By condition (1) of the definition, $R = (I:a) \in \mathcal{A}$ and so, by (2), $H \in \mathcal{A}$. This proves (3). If I and H both belong to $\mathcal{A}$ then for every $a \in H$, we have $((I \cap H):a) = (I:a) \cap (H:a) = (I:a) \in \mathcal{A}$ by (1), and so $I \cap H \in \mathcal{A}$ by (2), proving (4). □

(3.3) PROPOSITION: <u>The following conditions are equivalent for a set $\mathcal{A}$ of left ideals of</u> R:

(1) $\mathcal{A} = \mathcal{L}_\tau$ <u>for some</u> $\tau \in$ R-tors.

(2) $\mathcal{A}$ <u>is an idempotent filter</u>.

PROOF: (1) ⇒ (2): If $I \in \mathcal{L}_\tau$ then $R/I \in \mathcal{T}_\tau$, and so for any $a \in R$, we have $R/(I:a) = [Ra + I]/I \in \mathcal{T}_\tau$ by the closure of $\mathcal{T}_\tau$ under taking submodules. Therefore $(I:a) \in \mathcal{L}_\tau$.

If I is a left ideal of R and if there exists an $H \in \mathcal{L}_\tau$ with $(I:a) \in \mathcal{L}_\tau$ for every $a \in H$, consider the exact sequence

$$0 \to H/[I \cap H] \to R/I \to R/[I + H] \to 0$$

in R-mod. Then $R/[I + H]$ in τ-torsion since it is a homomorphic image of $R/H \in \mathcal{T}_\tau$, and $H/[I \cap H] \in \mathcal{T}_\tau$ since $a \in H$ implies that $(I \cap H:a) = (I:a) \in \mathcal{L}_\tau$, whence $[Ra + (I \cap H)]/[I \cap H] \in \mathcal{T}_\tau$. By the closure of $\mathcal{T}_\tau$ under extensions we then have $R/I \in \mathcal{T}_\tau$ or, in other words, $I \in \mathcal{L}_\tau$. Thus we have (2).

(2) ⇒ (1): If $\mathcal{A}$ satisfies (2), define the subfunctor F of the identity functor on R-mod by $F(M) = \{m \in M \mid (0:m) \in \mathcal{A}\}$. Then F is easily seen to be left exact. Furthermore, if $m + F(M) \in F(M/F(M))$ then $(F(M):m) \in \mathcal{A}$. But $(F(M):m) = \{r \in R \mid (0:rm) \in \mathcal{A}\} = \{r \in R \mid ((0:m):r) \in \mathcal{A}\}$. Since $(F(M):m)$ belongs to $\mathcal{A}$,

then by condition (2) of the definition of an idempotent filter we have $(0:m) \in \mathcal{A}$ and so $m \in F(M)$. Therefore $F(M/F(M)) = 0$. By Proposition 2.5 we therefore have $F = T_\tau$ for some $\tau \in$ R-tors and it is clear that in this case $\mathcal{A} = \mathcal{L}_\tau$. □

For any $\tau \in$ R-tors, set $L(\tau) = \cap \mathcal{L}_\tau$. Then $L(\tau)$ is a two-sided ideal of R. In general it is not true, however, that $L(\tau) \in \mathcal{L}_\tau$. Torsion theories satisfying this condition are very useful and will be studied in Section 22.

Let $\tau \in$ R-tors. We can endow each left R-module with a linear topological structure by defining a topology $X_\tau(M)$ on M a basis of neighborhoods of 0 in which is the family of all τ-dense submodules of M. If each left R-module is endowed with such a topology then every R-homomorphism $\alpha: M \to N$ becomes a continuous map of topological R-modules. In general, if N is a submodule of a left R-module M then the topology $X_\tau(N)$ need not coincide with the restriction of the topology $X_\tau(M)$ to N. However, we do have the following result.

(3.4) PROPOSITION: _If_ $\tau \in$ R-tors _and if_ N _is a_ τ-_dense submodule of a left_ R-_module_ M, _then_ $X_\tau(N)$ _coincides with the restriction of_ $X_\tau(M)$ _to_ N.

PROOF: If N' is a submodule of N, then from the short exact sequence $0 \to N/N' \to M/N' \to M/N \to 0$ we see that N' is τ-dense in N if and only if it is τ-dense in M. □

If $\tau \in R$-tors, then a submodule N of a left R-module M is said to be τ-_pure_ in M if and only if M/N is τ-torsion-free. The intersection of an arbitrary family of τ-pure submodules of a left R-module is again τ-pure. Indeed, if $\mathcal{A}$ is a family of τ-pure submodules of a left R-module M, then there exists a canonical monomorphism $M/\cap\mathcal{A} \to \Pi\{M/N \mid N \in \mathcal{A}\}$, the range of which is τ-torsion-free since $\mathcal{F}_\tau$ is closed under taking direct products and submodules. Therefore $\cap\mathcal{A}$ is τ-pure in M.

Any submodule N of a left R-module M is contained in at least one τ-pure submodule of M (namely M itself), and so there exists a unique minimal element of the family of all τ-pure submodules of M containing N. We call this the τ-_purification_ of N in M. If N' is the τ-purification of N in M then one easily sees that $N'/N = T_\tau(M/N)$. Moreover, N' is precisely the closure of N in the topology $X_\tau(M)$. In particular, $T_\tau(M)$ is the τ-purification of 0 in M.

If $\tau \in R$-tors then τ defines an operator $I \mapsto I^{p(\tau)}$ on the modular lattice of all left ideals of R, which assigns to each left ideal I its τ-purification $I^{p(\tau)}$ in R.

(3.5) PROPOSITION: <u>The correspondence</u> $\tau \mapsto p(\tau)$ <u>is monic</u>.

PROOF: Assume $p(\tau) = p(\tau')$. Then $\mathcal{F}_\tau = \{M \in R\text{-mod} \mid (0:m)^{p(\tau)} = (0:m) \text{ for every } m \in M\} = \{M \in R\text{-mod} \mid (0:m)^{p(\tau')} = (0:m) \text{ for every } m \in M\} = \mathcal{F}_{\tau'}$, and so $\tau = \tau'$ by Proposition 1.3. □

We see, therefore, that the operator $p(\tau)$ uniquely determines τ and we ask what operators on the lattice of all left ideal of R are of the form $p(\tau)$ for some $\tau \in R\text{-tors}$.

An operator $I \mapsto I^c$ on the modular lattice of all left ideals of R is called a <u>modular closure operator</u> if and only if the following conditions are satisfied:

(1) $I \subseteq I^c$ for all ${}_RI \subseteq R$;

(2) $I^c = I^{cc}$ for all ${}_RI \subseteq R$;

(3) $I_1 \subseteq I_2 \Rightarrow I_1^c \subseteq I_2^c$ for all $I_1, I_2 \subseteq R$;

(4) $(I^c:a) = (I:a)^c$ for all ${}_RI \subseteq R$ and $a \in R$.

(3.6) PROPOSITION: <u>The following conditions are equivalent for an operator</u> $I \mapsto I^c$ <u>on the modular lattice of all left ideals of</u> R:

(1) $c = p(\tau)$ <u>for some</u> $\tau \in R\text{-tors}$;

(2) c <u>is a modular closure operator</u>.

PROOF: (1) ⇒ (2): It is immediate from the definition of $p(\tau)$ that for every left ideal I of R, we get $I \subseteq I^{p(\tau)}$ and $I^{p(\tau)} = I^{p(\tau)p(\tau)}$. If $I \subseteq H$ are left ideals of R, then $I^{p(\tau)}/I$ is τ-torsion and so $[I^{p(\tau)} + H]/H$ is τ-torsion. Therefore, $I^{p(\tau)} \subseteq I^{p(\tau)} + H \subseteq H^{p(\tau)}$. Finally, if I is a left ideal of R and if $a \in R$, then $(I:a)^{p(\tau)} = \{r \in R \mid r + (I:a) \in T_\tau(R/(I:a))\} = \{r \in R \mid R/(0:r + (I:a))$ is τ-torsion$\} = \{r \in R \mid R/(I:ra)$ is τ-torsion$\} = \{r \in R \mid ra + I \in T_\tau(R/I)\} = (I^{p(\tau)}: a)$.

(2) ⇒ (1): Let $\mathcal{A} = \{_R I \subseteq R \mid I^c = R\}$. If $I \in \mathcal{A}$ and $a \in R$, then $(I:a)^c = (I^c:a) = (R:a) = R$ and so $(I:a) \in \mathcal{A}$. If I is a left ideal of R and if $H \in \mathcal{A}$ satisfies the condition that $(I:a) \in \mathcal{A}$ for every $a \in H$, then $R = (I:a)^c = (I^c:a)$ for every $a \in H$ implies that $H \subseteq I^c$. Thus $R = H^c \subseteq I^{cc} = I^c$, proving that $I \in \mathcal{A}$. Therefore $\mathcal{A}$ is an idempotent filter and so, by Proposition 3.3, $\mathcal{A} = \mathcal{L}_\tau$ for some $\tau \in R$-tors.

We are left, therefore, to show that $c = p(\tau)$ for this τ. Let I be a left ideal of R. If $a \in I^{p(\tau)}$ then $(I:a) \in \mathcal{L}_\tau$ and so $R = (I:a)^c = (I^c:a)$. Thus $a \in I^c$ and so $I^{p(\tau)} \subseteq I^c$. Conversely, if $a \in I^c$ then $R = (I^c:a) = (I:a)^c$ and so $(I:a) \in \mathcal{L}_\tau$ whence

$a + I \in T_\tau(R/I) = I^{p(\tau)}/I$. Therefore, $I^c \subseteq I^{p(\tau)}$ and so we have equality. □

(3.7) PROPOSITION: *If* I, H *are left ideals of* R *and if* $\tau \in R\text{-tors}$, *then* $I^{p(\tau)} \cap H^{p(\tau)} = (I \cap H)^{p(\tau)}$.

PROOF: Since $I \cap H \subseteq I$, $(I \cap H)^{p(\tau)}$ is contained in $I^{p(\tau)}$. Similarly, it is contained in $H^{p(\tau)}$ and hence in $I^{p(\tau)} \cap H^{p(\tau)}$.

If K_1 and K_2 are left ideals of R and if $a \in K_1 \cap K_2^{p(\tau)}$, then $((K_1 \cap K_2)^{p(\tau)} : a) = (K_1 \cap K_2 : a)^{p(\tau)} = ((K_1 : a) \cap (K_2 : a))^{p(\tau)} = (K_2 : a)^{p(\tau)} = (K_2^{p(\tau)} : a) = R$ and so $K_1 \cap K_2^{p(\tau)} \subseteq (K_1 \cap K_2)^{p(\tau)}$.

In particular, taking $K_1 = H$ and $K_2 = I$ we have $I^{p(\tau)} \cap H \subseteq (I \cap H)^{p(\tau)}$ so $(I^{p(\tau)} \cap H)^{p(\tau)} \subseteq (I \cap H)^{p(\tau)}$. Now taking $K_1 = I^{p(\tau)}$ and $K_2 = H$, we have $I^{p(\tau)} \cap H^{p(\tau)} \subseteq (I^{p(\tau)} \cap H)^{p(\tau)} \subseteq (I \cap H)^{p(\tau)}$, proving equality. □

Let $\mathcal{C}_\tau$ be the set of all τ-pure left ideals of R; that is to say, $I \in \mathcal{C}_\tau$ if and only if $I = I^{p(\tau)}$. Since $\mathcal{C}_\tau$ is closed under taking arbitrary intersections, we can define a complete lattice structure on $\mathcal{C}_\tau$ by setting $\wedge\mathcal{A} = \cap\mathcal{A}$ and $\vee\mathcal{A} = (\Sigma\mathcal{A})^{p(\tau)}$ for every subset $\mathcal{A}$ of $\mathcal{C}_\tau$.

(3.8) COROLLARY: *If* $\tau \in R\text{-tors}$, *then the complete lattice* $\mathcal{C}_\tau$ *is modular*.

PROOF: Note that the lattice of all left ideals of R is modular. If $H, I, K \in \mathcal{C}_\tau$ with $H \subseteq I$, $I \wedge (H \vee K) = I^{p(\tau)} \cap (H + K)^{p(\tau)} = [I \cap (H + K)]^{p(\tau)} = [H + (I \cap K)]^{p(\tau)} = H \vee (I \wedge K)$. □

Finally we have the following characterization of the τ-pure left ideals of R.

(3.9) PROPOSITION: *If* $E \in \tau \in R\text{-tors}$, *then a left ideal* I *of* R *is* τ*-pure in* R *if and only if* I *is the annihilator of a subset of* E.

PROOF: If A is a subset of E then $(0:A) = \{(0:x) \mid x \in A\}$. Since E is τ-torsion-free, each $(0:x)$ is τ-pure in R and so their intersection $(0:A)$ is τ-pure in R.

Conversely, let I be a τ-pure left ideal of R and let $A = \{x \in E \mid Ix = 0\}$. Clearly $I \subseteq (0:A)$. Let $a \in (0:A)$ and let $\alpha: R/(I:a) \to E$ be an R-homomorphism. Let $\beta: R/(I:a) \to R/I$ be the R-monomorphism defined by $\beta: r + (I:a) \mapsto ra + I$. By the injectivity of E there then exists an R-homomorphism $\alpha': R/I \to E$ for which $\alpha = \beta\alpha'$. Set $x = (1 + I)\alpha'$. Then $Ix = 0$ and so $x \in A$. But then also $ax = 0$ and so $\beta\alpha' = 0$ which implies that $\alpha = 0$.

Thus $\mathrm{Hom}_R(R/(I:a), E) = 0$ and so $(I:a) \in \mathcal{L}_\tau$. But I is τ-pure in R and so we must have $a + I = 0$. This implies that $(0:A) \subseteq I$ and so we have equality. □

References for Section 3

Chew [30]; Dickson [36]; Gabriel [52]; Goldman [63]; Hacque [67]; Lambek [90, 93]; Mishina and Skornyakov [104]; Sanderson [135]; Stenström [147]; Walker and Walker [164].

4. RELATIVE INJECTIVITY AND NEATNESS

If $\tau \in R$-tors we say that a left R-module M is τ-*injective* if and only if every diagram of the form

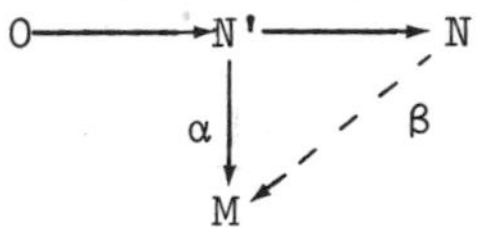

$(N/N' \in \mathcal{T}_\tau)$

can be completed commutatively. It is immediate from this definition that the class of all τ-injective left R-modules is closed under taking direct summands and finite direct sums.

(4.1) PROPOSITION: *Let* $\tau \in R$-tors. *For a left* R-*module* M *the following conditions are equivalent*:

(1) M *is* τ-*injective*.

(2) *Any diagram of the form*

$$\begin{array}{ccccc} 0 & \longrightarrow & I & \longrightarrow & R \\ & & {\scriptstyle\alpha}\downarrow & \swarrow {\scriptstyle\beta} & \\ & & M & & \end{array} \qquad (I \in \mathcal{L}_\tau)$$

can be completed commutatively.

(3) M *is* τ-*pure in* $E(M)$.

(4) $\mathrm{Ext}_R^1(N,M) = 0$ *for every* τ-*torsion left* R-*module* N.

(5) M *is a direct summand of a* τ-*pure submodule of any left* R-*module containing it*.

PROOF: (1) ⇒ (3): Let N be the τ-purification of M in $E(M)$. By (1) the identity map $M \to M$ can be extended to an R-homomorphism $\beta: N \to M$. Since M is large in $E(M)$, it is large in N and so β is a monomorphism. Since the restriction of β to M is the identity map, it follows that the inclusion $M \to N$ must be an epimorphism, i.e., $M = N$. Therefore, M is τ-pure in $E(M)$.

(3) ⇒ (2): Let $I \in \mathcal{L}_\tau$ and let $\alpha: I \to M$ be an R-homomorphism. By the injectivity of $E(M)$ there exists an R-homomorphism φ making the diagram

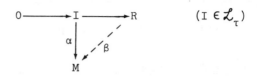

$$\begin{array}{ccccc} 0 & \longrightarrow & I & \longrightarrow & R \\ & & \Big\downarrow \alpha & & \Big\downarrow \varphi \\ 0 & \longrightarrow & M & \longrightarrow & E(M) \end{array}$$

commute. Set $x = 1\varphi$. Then $[Rx + M]/M \cong R/(M:x)$ which is τ-torsion since $I \subseteq (M:x)$. But by (3) this means that $x \in M$. Therefore $R\varphi \subseteq M$, proving (2).

(2) $\Rightarrow$ (1): We adapt the classical argument of Baer. Let N' be a τ-dense submodule of a left R-module N and let $\alpha\colon N' \to M$ be an R-homomorphism. Consider the set of all pairs (W,β), where W is a submodule of N containing N' and $\beta\colon W \to M$ is an R-homomorphism extending α. Order this set by putting $(W,\beta) \leq (W', \beta')$ if and only if $W \subseteq W'$ and β is the restriction of β' to W. Then this set is inductive and so by Zorn's Lemma it has a maximal element (W_0, β_0). We claim that $W_0 = N$, in which case the proof will be completed. Assume $W_0 \neq N$ and pick $x \in N \setminus W_0$. Set $I = (W_0:x)$. Then $(N':x) \subseteq I$ and so $I \in \mathcal{L}_\tau$. Consider the R-homomorphism $\varphi\colon I \to M$ defined by $\varphi\colon a \mapsto (ax)\beta_0$. By (2), φ can be extended to an R-homomorphism $\psi\colon R \to M$. Now define $\beta_1\colon W_0 + Rx \to M$ by $w_0 + rx \mapsto w_0\beta_0 + r\psi$. Then this is a well-defined R-homomorphism which properly extends β_0, contradicting the maximality of (W_0, β_0). Thus we must have $W_0 = N$.

(3) ⇒ (4): By (3), $E(M)/M$ is τ-torsion-free. From the exact sequence $0 \to M \to E(M) \to E(M)/M \to 0$ we therefore obtain, for every τ-torsion left R-module N, an exact sequence $\mathrm{Hom}_R(N,E(M)/M) \to \mathrm{Ext}^1_R(N,M) \to \mathrm{Ext}^1_R(N,E(M))$, the left-hand end of which is 0 since N is τ-torsion and $E(M)/M$ is τ-torsion-free, while the right-hand end is 0 since $E(M)$ is injective. Therefore $\mathrm{Ext}^1_R(N,M) = 0$.

(4) ⇒ (1): This proof is trivial.

(3) ⇒ (5): Let $M \subseteq M' \in$ R-mod. Then $E(M') = E(M) \oplus N$ for some submodule N of $E(M')$. If $N' = M' \cap N$ then $M'/[M \oplus N']$ is isomorphic to a submodule of $E(M')/[M \oplus N]$ and furthermore $E(M')/[M \oplus N] \cong [E(M) \oplus N]/[M \oplus N] \cong E(M)/M$ which is τ-torsion-free by (3). Therefore $M \oplus N'$ is τ-pure in M'.

(5) ⇒ (3): By (5), there exists a submodule N of $E(M)$ such that $M \oplus N$ is τ-pure in $E(M)$. But M is large in $E(M)$ and so we must have $N = 0$, which proves (3). □

(4.2) PROPOSITION: *Let* $\tau \in$ R-tors. *For a* τ-*torsion left* R-*module* M *the following conditions are equivalent*:

(1) M *is* τ-*injective*.

(2) $M = T_\tau(E(M))$.

(3) (i) M *is quasi-injective; and*

(ii) *If* $I \in \mathcal{L}_\tau$ *and if* I' *is a left ideal*

<u>of</u> R <u>contained in</u> I <u>for which</u> I/I' <u>is isomorphic to a submodule of</u> M, <u>then</u> $I' = I \cap (0:m)$ <u>for some</u> $m \in M$.

PROOF: (1) ⇒ (2): By Proposition 4.1, M is τ-pure in E(M) and so $T_\tau(E(M)) \subseteq M$. But M is also τ-torsion and so the reverse containment holds, proving equality.

(2) ⇒ (3): By (2), M is a fully invariant submodule of an injective left R-module and so is quasi-injective, proving (i). As for (ii), if I/I' is isomorphic to a submodule of M and $I \in \mathcal{L}_\tau$, then there exists an R-homomorphism $\alpha: I \to M$ with kernel I'. By (1), α can then be extended to an R-homomorphism $\beta: R \to M$. Then $\ker(\beta) = (0:1\beta)$ and $I' = I \cap (0:1\beta)$.

(3) ⇒ (1): Let $I \in \mathcal{L}_\tau$ and let $\alpha: I \to M$ be an R-homomorphism if $I' = \ker(\alpha)$ then I/I' is isomorphic to a submodule of M, and so by (3) there exists an $m \in M$ with $I' = I \cap (0:m)$. Then α clearly has an extension $\alpha': I + (0:m) \to M$, the kernel of which is (0:m). By [49, Lemma 2] the quasi-injectivity of M implies that α' can be extended to an R-homomorphism $R \to M$, proving (1). □

Let $\tau \in$ R-tors. By Proposition 4.1 we see that for any left R-module M, the τ-purification of M in E(M) is a τ-injective submodule of E(M) containing M as a large submodule, and indeed is the minimal such submodule of $E(M)$. We therefore call the τ-purification of M in $E(M)$ the τ-<u>injective hull</u> of M and denote it by $E_\tau(M)$. As we shall see later, $E_\tau(M)$ plays an important part in the construction of localizations.

Let $\tau \in R$-tors. An R-homomorphism $\varphi: M' \to M$ is said to be τ-<u>neat</u> if and only if the following conditions are equivalent for every τ-dense submodule N' of a left R-module N and for every R-homomorphism $\alpha: N' \to M'$:

(1) There exists a submodule W of N properly containing N' and an R-homomorphism $\beta: W \to M$ making the following diagram commute.

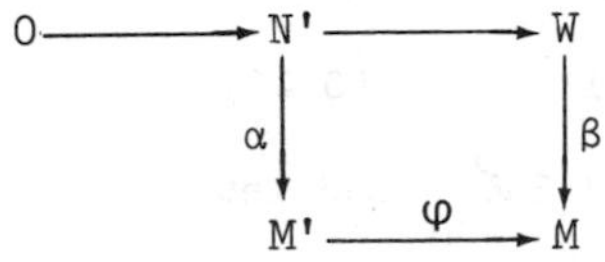

(2) There exists a submodule W' of N properly containing N' and an R-homomorphism $\beta': W' \to M'$ making the following diagram commute.

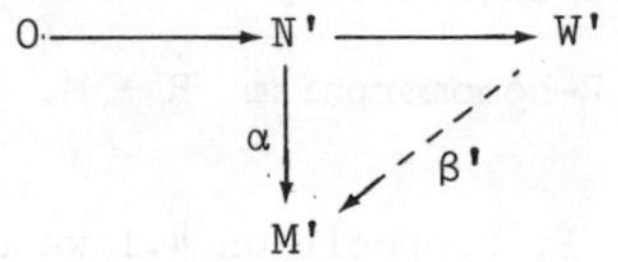

A variation on the argument used in the proof of

Proposition 4.1 shows that it suffices to consider the case $N = R$.

(4.3) PROPOSITION: *If* $\tau \in R\text{-tors}$ *then the following conditions exist.*

(1) *If* $\alpha: M' \to M$ *and* $\beta: M \to M''$ *are* τ-*neat* R-*homomorphisms, then* $\alpha\beta$ *is* τ-*neat*.

(2) *If* $\alpha: M' \to M$ *and* $\beta: M \to M''$ *are* R-*homomorphisms with* $\alpha\beta$ τ-*neat, then* α *is* τ-*neat*.

(3) *A left* R-*module* M *is* τ-*injective if and only if the zero map* $M \to 0$ *is* τ-*neat*.

PROOF: The proof follows directly from the definition. □

(4.4) PROPOSITION: *If* $\{\varphi_i: M_i' \to M_i \mid i \leq i \leq n\}$ *are* τ-*neat* R-*homomorphisms for some* $\tau \in R\text{-tors}$ *and if* $M = \oplus M_i$, $M' = \oplus M_i'$, *and* $\varphi = \oplus \varphi_i: M' \to M$, *then* φ *is* τ-*neat*.

PROOF: It suffices to prove this for the case $n = 2$ and then to proceed by induction. For $i = 1, 2$ let $\pi_i': M' \to M_i'$ and $\pi_i: M \to M_i$ be the canonical projections. Let $I' \in \mathcal{L}_\tau$ and let $\alpha \in \mathrm{Hom}_R(I', M')$. Let I be a left ideal of R properly containing I' and let $\beta: I \to M$ be an R-homomorphism making the diagram

$$\begin{array}{ccccc} 0 & \longrightarrow & I' & \longrightarrow & I \\ & & {\scriptstyle\alpha}\downarrow & & \downarrow{\scriptstyle\beta} \\ & & M' & \xrightarrow{\varphi} & M \end{array}$$

commute. Then the diagram

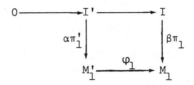

$$\begin{array}{ccccc} 0 & \longrightarrow & I' & \longrightarrow & I \\ & & {\scriptstyle\alpha\pi_1'}\downarrow & & \downarrow{\scriptstyle\beta\pi_1} \\ & & M_1' & \xrightarrow{\varphi_1} & M_1 \end{array}$$

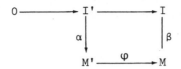

commutes and so by the τ-neatness of φ_1 there exists a left ideal I_1 of R contained in I and properly containing I' and an R-homomorphism $\beta_1: I_1 \to M_1'$ making the diagram

$$\begin{array}{ccccc} 0 & \longrightarrow & I' & \longrightarrow & I_1 \\ & & {\scriptstyle\alpha}\downarrow & & \downarrow{\scriptstyle\beta_1} \\ & & M' & \xrightarrow{\pi_1'} & M_1' \end{array}$$

commute. Furthermore, the diagram

$$\begin{array}{ccccc} 0 & \longrightarrow & I' & \longrightarrow & I_1 \\ & & {\scriptstyle\alpha\pi_2'}\downarrow & & \downarrow{\scriptstyle\beta|I_1\pi_2} \\ & & M_2' & \xrightarrow{\varphi_2} & M_2 \end{array}$$

commutes and so there exists a left ideal I_2 of R contained in I_1 and properly containing I' and an

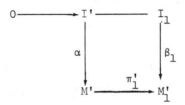

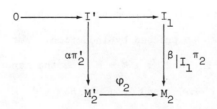

R-homomorphism $\beta_2: I_2 \to M'_2$ making the diagram

$$\begin{array}{ccccc} 0 & \longrightarrow & I' & \longrightarrow & I_2 \\ & & \downarrow{\scriptstyle\alpha} & & \downarrow{\scriptstyle\beta_2} \\ & & M' & \xrightarrow{\pi'_2} & M'_2 \end{array}$$

commute. Define $\beta': I_2 \to M$ by $\beta': a \mapsto a\beta_1 + a\beta_2$. Then β' extends α, proving the τ-neatness of φ. □

(4.5) PROPOSITION: Let $\tau \in$ R-tors and let

$$\begin{array}{ccccc} 0 & \longrightarrow & N' & \xrightarrow{\zeta'} & M' \\ & & \downarrow{\scriptstyle\alpha'} & & \downarrow{\scriptstyle\alpha} \\ & & M & \xrightarrow{\zeta} & N \end{array}$$

be a pullback diagram with $N \in \mathcal{F}_\tau$. Then α' is τ-neat whenever α is.

PROOF: Let $I \in \mathcal{L}_\tau$ and let I' be a left ideal of R properly containing I. Assume that we have R-homomorphisms $\psi: I \to N'$ and $\theta: I' \to M$ satisfying $\psi\alpha' = \lambda\theta$, where $\lambda: I \to I'$ is the canonical inclusion. By the τ-neatness of α there then exists a left ideal H of R contained in I' and properly containing I and an R-homomorphism $\varphi: H \to M'$ satisfying $\psi\zeta' = \lambda'\varphi$, where $\lambda': I \to H$ is the canonical inclusion.

We claim that $\varphi\alpha = \theta_{|H}\zeta$. Indeed, if $\omega = \varphi\alpha - \theta_{|H}\zeta$

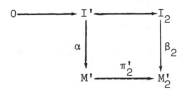

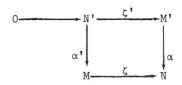

then $\lambda'\omega = \lambda'\varphi\alpha - \lambda'\theta|_H\zeta = \psi\zeta'\alpha - \lambda'\theta|_H\zeta = 0$, and so ω induces an R-homomorphism $H/I \to N$. But $H/I \in \mathcal{T}_\tau$ and $N \in \mathcal{F}_\tau$ and so $\bar{\omega} = 0$ whence $\omega = 0$.

By the definition of a pullback we then have an R-homomorphism ξ making the diagram

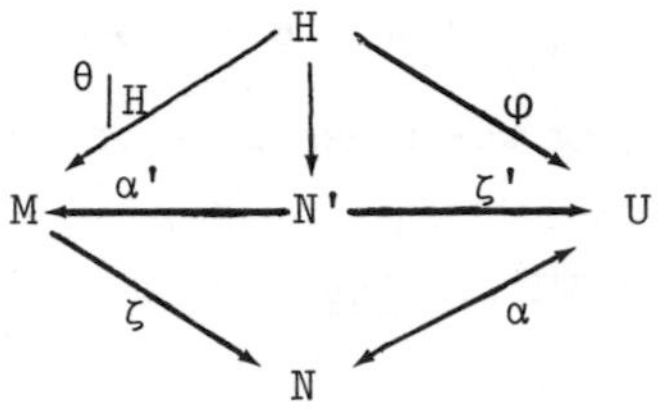

commute. In particular, $\psi\zeta' = \lambda'\varphi = \lambda'\xi\zeta'$. Since ζ' is a monomorphism, this implies that $\psi = \lambda'\xi$ and so α' is τ-neat. □

(4.6) PROPOSITION: *The following conditions are equivalent for* $\tau \in$ R-tors:

(1) *Every* $I \in \mathcal{L}_\tau$ *is projective.*

(2) *Homomorphic images of* τ-*injective left R-modules are* τ-*injective.*

(3) *If* α *and* β *are R-epimorphisms for which* $\alpha\beta$ *is defined, then* $\alpha\beta$ *is* τ-*neat if and only if both* α *and* β *are* τ-*neat.*

PROOF: (1) ⇒ (3): If both α and β are τ-neat, then $\alpha\beta$ is τ-neat by Proposition 4.3 (1). Conversely, assume that $\alpha\beta$ is τ-neat. Then α is τ-neat by Proposition 4.3(2). For $I \in \mathcal{L}_\tau$ let H be a left ideal of R properly containing I and consider a diagram of the form

$$\begin{array}{ccccc} 0 & \longrightarrow & I & \xrightarrow{\lambda} & H \\ & & \downarrow{\scriptstyle\psi} & & \downarrow{\scriptstyle\theta} \\ M' & \xrightarrow{\alpha} & M & \xrightarrow{\beta} & M'' \end{array}$$

Since I is projective, there exists an R-homomorphism $\zeta\colon I \to M'$ satisfying $\zeta\alpha = \psi$. Since $\alpha\beta$ is τ-neat, there exists a submodule H' of H properly containing I and an R-homomorphism $\xi\colon H' \to M'$ satisfying $\lambda\xi = \zeta$. Therefore $\xi\alpha\colon H \to M$ extends ψ, proving that β is τ-neat.

(3) ⇒ (2): Let M be a τ-injective left R-module and let $M \overset{\alpha}{\to} N \overset{\beta}{\to} 0$ be exact. By Proposition 4.3(3), $\alpha\beta$ is τ-neat and so, by (3), β is τ-neat. By Proposition 4.3(3), this proves that N is τ-injective.

(2) ⇒ (1): Let $I \in \mathcal{L}_\tau$ and let $\alpha\colon M \to N$ be an R-epimorphism with M τ-injective. By (2), N is also τ-injective. Let $\beta \in \mathrm{Hom}_R(I,N)$. By the τ-injectivity of N, there exists an R-homomorphism $\varphi\colon R \to N$ extending β. By the projectivity of R there exists an R-homomorphism $\psi\colon R \to M$ satisfying $\varphi = \psi\alpha$. Then $\beta = \psi|_I\alpha$ and so I is projective with respect to α.

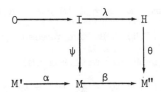

Now let M be an arbitrary left R-module and let $\alpha: M \to N$ be an R-epimorphism with kernel K. Let ν be the canonical R-epimorphism $E_\tau(M) \to E_\tau(M)/K\lambda$, where $\lambda: M \to E_\tau(M)$ is the inclusion map. Since $N \cong M/K$ there exists an R-monomorphism $\omega: N \to E_\tau(M)/K\lambda$ satisfying $\alpha\omega = \lambda\nu$. Let $\beta \in \mathrm{Hom}_R(I,N)$. By the preceding paragraph, there exists an R-homomorphism $\psi: I \to E_\tau(M)$ satisfying $\psi\nu = \beta\omega$. Moreover, $\mathrm{im}(\psi) \subseteq M\lambda + \ker(\nu) = M\lambda + K\lambda \subseteq M\lambda$. Therefore $\psi' = \psi\lambda^{-1}$ is a well-defined R-homomorphism $I \to M$ satisfying $\psi'\alpha = \beta$. This proves that I is projective. □

Infinite direct sums of τ-neat R-homomorphisms need not be τ-neat. See Section 14 for a characterization of those torsion theories τ for which this property holds.

A submodule M' of a left R-module M is said to be τ-<u>neat</u> in M if and only if the inclusion map $M' \to M$ is τ-neat.

(4.7) PROPOSITION: <u>Let</u> $\tau \in$ R-tors <u>and let</u> M <u>be a</u> τ-<u>torsion-free left</u> R-<u>module</u>. <u>Then a submodule</u> M' <u>of</u> M <u>is</u> τ-<u>neat in</u> M <u>if and only if it is</u> τ-<u>pure in</u> M.

PROOF: Let M' be a submodule of the τ-torsion-free left R-module M. Assume that M' is τ-neat in M but not τ-pure. Let W be the τ-purification of M' in M.

Then we have the commutative diagram

$$\begin{array}{ccccc} 0 & \longrightarrow & M' & \xrightarrow{\zeta} & W \\ & & = \big\downarrow & & \big\downarrow \\ 0 & \longrightarrow & M' & \longrightarrow & M \end{array}$$

(all maps inclusions). By τ-neatness there exists a submodule W' of W properly containing M' and an R-homomorphism $\beta: W' \to M'$ such that $\zeta\beta$ is the identity on M'. Then W' is τ-torsion-free since it is a submodule of M. Moreover, the existence of β shows that M' is a direct summand of W' so that W'/M' is also τ-torsion-free, a contradiction.

Conversely, let M' be a τ-pure submodule of M. Consider a commutative diagram of the form

$$\begin{array}{ccccc} 0 & \longrightarrow & N' & \longrightarrow & N \\ & & \alpha \big\downarrow & & \big\downarrow \beta \\ 0 & \longrightarrow & M' & \longrightarrow & M \end{array}$$

where N' is a τ-dense submodule of N. Then β induces an R-homomorphism $\overline{\beta}: N/N' \to M/M'$ defined by $\overline{\beta}: x + N' \mapsto x\beta + M'$. But N/N' is τ-torsion and M/M' is τ-torsion-free so we must have $\overline{\beta} = 0$, which implies that $N\beta \subseteq M'$. This suffices to show that M' is τ-neat in M. □

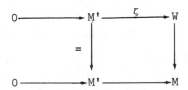

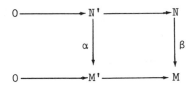

(4.8) PROPOSITION: *Let* $\tau \in$ R-tors *and let* $\varphi: M \to N$ *be a* τ-*neat* R-*homomorphism. If* $M \in \mathcal{F}_\tau$ *then the following conditions are equivalent for a submodule* M' *of* $\ker(\varphi)$:

(1) M' *is* τ-*injective*;

(2) M' *is* τ-*pure in* M.

PROOF: (1) ⇒ (2): Let M'' be the τ-purification of M' in M. By the τ-injectivity of M', the identity map $M' \to M'$ extends to an R-homomorphism $M'' \to M'$ and so the exact sequence $0 \to M' \to M'' \to M''/M' \to 0$ splits. Thus M''/M' is isomorphic to a submodule of $M'' \subseteq M$ and so is τ-torsion-free, a contradiction unless $M'' = M'$.

(2) ⇒ (1): Let $I \in \mathcal{L}_\tau$ and let $\alpha \in \text{Hom}_R(I,M')$. By an argument similar to that in the proof of (2) ⇒ (1) of Proposition 4.1, there exists a maximal extension $\alpha': I' \to M'$ of α, where $I \subseteq {}_R I' \subseteq R$. By Proposition 4.1 it suffices to prove that $I' = R$.

If $I' \neq R$ then the zero map $R \to N$ properly extends $\alpha'\varphi$, and so by the τ-neatness of φ there exists a left ideal H of R properly containing I' and an R-homomorphism $\beta: H \to M$ extending α'. By the maximality of I', $H\beta \not\subseteq M'$ and so $0 \neq [H\beta + M']/M' \subseteq M/M'$. Since M' is τ-pure in M, it is τ-pure in $H\beta + M'$. But by the maximality of I', $[H\beta + M']/M' \cong H\beta/[H\beta \cap M'] = H\beta/I'\beta$

which is τ-torsion since $H/I' \subseteq R/I' \in \mathcal{T}_\tau$. Therefore we have a contradiction, proving that $I' = R$. □

References for Section 4

Bowe [20]; Fuchs [49, 50]; Gabriel [52]; Golan and Teply [47, 48]; Goldman [63]; Helzer [71]; Lambek [90]; Maranda [98]; Mishina and Skornyakov [104]; Sanderson [135]; Stenström [147]; Walker and Walker [164].

5. ABSOLUTE PURITY

Let $\tau \in R$-tors. If a left R-module M is τ-torsion-free and τ-injective we say that M is *absolutely* τ-*pure*. This terminology is justified by the following result.

(5.1) PROPOSITION: *Let* $\tau \in R$-tors. *For a left* R-*module* M *the following conditions are equivalent*:

(1) M *is absolutely* τ-*pure*.

(2) M *is* τ-*torsion-free and is a* τ-*pure submodule of every* τ-*torsion-free left* R-*module containing it*.

(3) *Any diagram of the form*

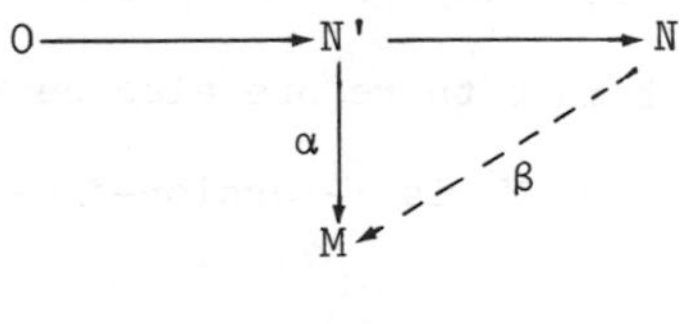

$(N/N' \in \mathcal{T}_\tau)$

can be completed commutatively in a

unique manner.

PROOF: (1) ⇔ (2): Assume that M is absolutely τ-pure and let M' be a τ-torsion-free left R-module containing M. By Proposition 4.1(5) there exists submodule N of M' such that $M \oplus N$ is τ-pure in M'. Therefore we have an exact sequence $0 \to [M \oplus N]/M \to M'/M \to M'/[M \oplus N] \to 0$. Moreover, $[M \oplus N]/M \cong N$ which is τ-torsion-free since M' is τ-torsion-free. Therefore M'/M is τ-torsion-free and so M is τ-pure in M'.

Conversely, if M is τ-pure in any τ-torsion-free left R-module containing it, then it is in particular τ-pure in $E(M)$ and so is τ-injective by Proposition 4.1.

(1) ⇔ (3): Let M be τ-torsion-free and consider the diagram in (3). If there are two R-homomorphisms β and β' making the diagram commute, then $\beta-\beta'$ induces an R-homomorphism $N/N' \to M$. But N/N' is τ-torsion and M is τ-torsion-free, so this must be the zero map. Therefore $\beta = \beta'$ and so α has a unique extension.

Conversely, (3) clearly implies that M is τ-injective. Moreover, by (3) the zero map $0 \to M$ has a unique extension to $T_\tau(M) \to M$, which must therefore also be the zero map. Thus $T_\tau(M) = 0$ and so M is τ-torsion-free. □

Let $\tau \in$ R-tors. We denote by $\mathcal{E}_\tau$ the class of all absolutely τ-pure left R-modules.

(5.2) PROPOSITION: The correspondence $\tau \mapsto \mathcal{E}_\tau$ is monic.

PROOF: Assume that $\mathcal{E}_\tau = \mathcal{E}_{\tau'}$. Then $\mathcal{F}_\tau = \{M \in \text{R-mod} \mid E(M) \text{ is absolutely } \tau\text{-pure}\} = \{M \in \text{R-mod} \mid E(M) \text{ is absolutely } \tau'\text{-pure}\} = \mathcal{F}_{\tau'}$, and so $\tau = \tau'$ by Proposition 1.3. □

It is possible to characterize those full subcategories of R-mod, the object class of which is precisely $\mathcal{E}_\tau$ for some $\tau \in$ R-tors. Since we are trying to keep the amount of category theory in this volume to a minimum, we shall not do so here. The interested reader is referred to [147].

It is also possible to weaken the requirement of τ-injectivity.

(5.3) PROPOSITION: Let $\tau \in$ R-tors. Then the following conditions are equivalent for a τ-torsion-free left R-module M:

(1) M is absolutely τ-pure.

(2) There exists a τ-dense submodule N of M such that for all $I \in \mathcal{L}_\tau$, every diagram of the form

$$\begin{array}{ccccc} 0 & & \to I & \longrightarrow & R \\ & & \alpha \downarrow & & \downarrow \\ 0 & \longrightarrow & N & \longrightarrow & M \end{array}$$

can be completed commutatively.

PROOF: (1) ⇒ (2) follows immediately by taking $N = M$.

(2) ⇒ (1): Let $I \in \mathcal{L}_\tau$ and let $\alpha: I \to M$ be an R-homomorphism. Set $H = (N \cap I\alpha)\alpha^{-1} \subseteq I$. Then I/H is isomorphic to a submodule of M/N and so is τ-torsion. In particular, from the exactness of the sequence $0 \to I/H \to R/H \to R/I \to 0$ and the closure of $\mathcal{T}_\tau$ under extensions, we have that $H \in \mathcal{L}_\tau$.

If α' is the restriction of α to H, then by (2) there exists an R-homomorphism $\beta: R \to M$ extending α'. Moreover, $H(\alpha - \beta) = 0$, and so $\alpha - \beta$ induces an R-homomorphism $I/H \to M$. But I/H is τ-torsion and M is τ-torsion-free, so this must be the zero map. Therefore β extends α. This proves that M is τ-injective and so absolutely τ-pure. □

(5.4) PROPOSITION: *If* $\tau \in R$-tors *and* $M \in \mathcal{E}_\tau$ *then the following conditions are equivalent for a submodule* N *of* M:

(1) N *is absolutely* τ-*pure*.

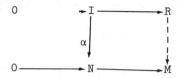

(2) N <u>is</u> τ-<u>pure in</u> M.

(3) N <u>is</u> τ-<u>injective</u>.

PROOF: (1) $\Leftrightarrow$ (3): The proof is trivial.

(2) $\Leftrightarrow$ (3): This follows from Propositions 4.8 and 4.3(3). □

(5.5) PROPOSITION: <u>The following conditions are equivalent for</u> $\tau \in$ R-tors:

(1) <u>The functor</u> $T_\tau(\_)$ <u>is exact</u>.

(2) $\mathcal{E}_\tau = \mathcal{F}_\tau$.

(3) $\mathcal{F}_\tau$ <u>is closed under taking homomorphic images</u>.

(4) <u>There exists a</u> $\tau' \in$ R-tors <u>with</u> $\mathcal{F}_\tau = \mathcal{T}_{\tau'}$.

(5) <u>For all</u> $I \in \mathcal{L}_\tau$, <u>we have</u> $I + T_\tau(R) = R$.

PROOF: (1) $\Rightarrow$ (2): Let M, N be τ-torsion-free left R-modules with $M \subseteq N$. Applying the exact functor $T_\tau(\_)$ to the exact sequence $0 \to M \to N \to N/M \to 0$, we deduce that $T_\tau(N/M) = 0$ and so M is τ-pure in N. By Proposition 5.4, M is therefore absolutely τ-pure. This suffices to prove (2).

(2) $\Rightarrow$ (3): If $M \in \mathcal{F}_\tau$ and if N is a submodule of M, then by (2) N is absolutely τ-pure and so $M/N \in \mathcal{F}_\tau$. This proves (3).

(3) $\Rightarrow$ (4): This follows from Propositions 1.5 and 1.6.

(3) $\Rightarrow$ (5): Let $I \in \mathcal{L}_\tau$ and let $H = I + T_\tau(R)$. Then R/H is a homomorphic image of $R/I \in \mathcal{T}_\tau$ and so is τ-torsion.

On the other hand, R/H is a homomorphic image of $R/T_\tau(R) \in \mathcal{F}_\tau$ and so, by (3), is τ-torsion-free. Therefore $R/H = 0$ and so $H = R$.

(5) $\Rightarrow$ (1): Let N be a submodule of a left R-module M and let $x + N \in T_\tau(M/N)$. Then $I = (N:x) \in \mathcal{L}_\tau$. By (5), $I + T_\tau(R) = R$ and so there exist $a \in I$ and $b \in T_\tau(R)$ such that $a + b = 1$ whence $x = 1x = ax + bx \in N + T_\tau(M)$. Therefore $T_\tau(M/N) \subseteq [N + T_\tau(M)]/N$. Conversely, $[N + T_\tau(M)]/N$ is a homomorphic image of $T_\tau(M)$ and so is τ-torsion. Thus $T_\tau(M/N) = [N + T_\tau(M)]/N \cong T_\tau(M)/[N \cap T_\tau(M)] = T_\tau(M)/T_\tau(N)$, proving (1). □

If $\tau \in$ R-tors satisfies the conditions of Proposition 5.5, then we clearly see that a left ideal I of R is τ-pure in R if and only if $I \supseteq T_\tau(R)$.

(5.6) PROPOSITION: _If_ $\tau \in$ R-tors _and_ $\mathcal{F}_\tau$ _is closed under taking homomorphic images, and if_ $\alpha: M \to N$ _is an R-homomorphism the kernel of which is small, then_ $N \in \mathcal{T}_\tau$ _implies_ $M \in \mathcal{T}_\tau$.

PROOF: Let $K = \ker(\alpha)$. If N is τ-torsion then so is $\mathrm{im}(\alpha)$ and hence so is M/K. Therefore $M/[K + T_\tau(M)]$ is τ-torsion. On the other hand, $M/[K + T_\tau(M)]$ is a homomorphic image of the τ-torsion-free left R-module $M/T_\tau(M)$ and so, by hypothesis, is τ-torsion-free. Therefore, by the smallness of K, we have $M = K + T_\tau(M) = T_\tau(M)$ and so M is τ-torsion. □

(5.7) PROPOSITION: *Let* $\tau \in R\text{-tors}$ *and assume that every cyclic* τ*-torsion left* R*-module has a projective cover. Then the following conditions are equivalent:*

(1) $\mathcal{F}_\tau$ *is closed under taking homomorphic images.*

(2) $\mathcal{T}_\tau$ *is closed under taking projective covers of cyclic left* R*-modules.*

PROOF: (1) ⇒ (2) follows directly from Proposition 5.6. Conversely, assume (2) and let $\alpha: N \to N'$ be an R-epimorphism with $N \in \mathcal{F}_\tau$. If M is a cyclic τ-torsion module and if $\mu: P \to M$ is a projective cover of M, then P is τ-torsion and so $\mathrm{Hom}_R(P,N) = 0$. By the exactness of the sequence $\mathrm{Hom}_R(P,N) \to \mathrm{Hom}_R(P,N') \to 0$ we deduce that $\mathrm{Hom}_R(P,N') = 0$. Therefore, the exactness of the sequence $0 \to \mathrm{Hom}_R(M,N') \to \mathrm{Hom}_R(P,N')$ implies that $\mathrm{Hom}_R(M,N') = 0$. This is true then for every cyclic τ-torsion left R-module M. Since $\mathcal{T}_\tau$ is closed under taking submodules, this implies that $T_\tau(N') = 0$ and so $N' \in \mathcal{F}_\tau$. □

References for Section 5

Alin and Dickson [5]; Beachy [10]; Bland [19]; Chamard [26]; Gabriel [52]; Goldman [63]; Helzer [71]; Lambek [90]; Maranda [98]; Rubin [132]; Stenström [147]; Teply [154]; Walker and Walker [164].

6. LOCALIZATION

Let $\tau \in$ R-tors. We can define a functor Q_τ : R-mod $\to$ R-mod as follows:

(1) If M is a left R-module, set $Q_\tau(M) = E_\tau(M/T_\tau(M))$.

(2) If $\alpha\colon M \to N$ is an R-homomorphism, then $T_\tau(M)\alpha \subseteq T_\tau(N)$ by Proposition 2.2 and so α induces an R-homomorphism $\bar{\alpha}\colon M/T_\tau(M) \to N/T_\tau(N)$. By τ-injectivity there then exists an R-homomorphism β making the diagram

$$\begin{array}{ccccc}
0 & \longrightarrow & M/T_\tau(M) & \longrightarrow & E_\tau(M/T_\tau(M)) \\
 & & \big\downarrow \bar{\alpha} & & \big\downarrow \beta \\
0 & \longrightarrow & N/T_\tau(N) & \longrightarrow & E_\tau(N/T_\tau(N))
\end{array}$$

commute, and furthermore this β is unique by Proposition 5.1. Then set $Q_\tau(\alpha) = \beta$.

It is straightforward to check that $Q_\tau(\_)$ is indeed an endofunctor of R-mod. We call $Q_\tau(\_)$ the τ-<u>localization functor</u> on R-mod. Furthermore, we have a natural transformation $\hat{\tau}$: identity on R-mod $\to Q_\tau(\_)$ which then defines, for every left R-module M, an R-homomorphism $\hat{\tau}_M\colon M \to Q_\tau(M)$. Indeed, $\hat{\tau}_M$ is just the composition of the canonical projection $M \to M/T_\tau(M)$ and the canonical embedding $M/T_\tau(M) \to Q_\tau(M)$. We immediately see that

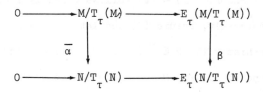

$\ker(\hat{\tau}_M) = T_\tau(M)$, and so $\hat{\tau}_M$ is a monomorphism if and only if M is τ-torsion-free, and is the zero map if and only if M is τ-torsion.

(6.1) PROPOSITION: *For* $\tau \in$ R-tors, *the endofunctor* $Q_\tau(\_)$ *of* R-mod *is idempotent and left exact.*

PROOF: That $Q_\tau(Q_\tau(M)) = Q_\tau(M)$ for every $M \in$ R-mod follows directly from the definition. Now let $\alpha: M \to N$ be an R-monomorphism. Then α induces an R-homomorphism $\bar{\alpha}: M/T_\tau(M) \to N/T_\tau(N)$ which is also a monomorphism by the closure of $\mathcal{J}_\tau$ under taking submodules and extensions. Since $M/T_\tau(M)$ is large in $Q_\tau(M)$, this implies that $Q_\tau(\alpha)$ is also a monomorphism. From here the left exactness of $Q_\tau(\_)$ follows directly. □

(6.2) PROPOSITION: *Let* $\tau \in$ R-tors. *Then the following conditions are equivalent for an* R-*monomorphism* $\alpha: M \to N$:

(1) $M\alpha$ *is* τ-*dense in* N.

(2) $Q_\tau(\alpha)$ *is an isomorphism*.

PROOF: We have an exact sequence $0 \to M \overset{\alpha}{\to} N \overset{\nu}{\to} N/M\alpha \to 0$ and so by Proposition 6.1 we have an exact sequence $0 \to Q_\tau(M) \to Q_\tau(N) \to Q_\tau(N/M\alpha)$.

(1) ⇒ (2): By (1), $Q_\tau(N/M\alpha) = 0$ since $N/M\alpha$ is

τ-injective and so $Q_\tau(\alpha)$ is an isomorphism.

(2) $\Rightarrow$ (1): By the definition of $Q_\tau(\_)$ we have the commutative diagram

$$\begin{array}{ccccc} N & \longrightarrow & N/T_\tau(N) & \xrightarrow{\lambda} & Q_\tau(N) \\ \downarrow{\scriptstyle \nu} & & \downarrow{\scriptstyle \bar{\nu}} & & \downarrow{\scriptstyle Q_\tau(\nu)} \\ N/M\alpha & \longrightarrow & [N/M\alpha]/T_\tau(N/M\alpha) & \xrightarrow{\lambda'} & Q_\tau(N/M\alpha) \end{array}$$

where λ and λ' are monomorphisms. By (2), $Q_\tau(\nu) = 0$ and so $\bar{\nu} = 0$. Since ν is an epimorphism, so is $\bar{\nu}$ and so $N/M\alpha = T_\tau(N/M\alpha)$, proving that $M\alpha$ is τ-dense in M. □

We now want to characterize $Q_\tau(M)$, for a given left R-module M, at least up to isomorphism.

(6.3) PROPOSITION: *For* $\tau \in$ R-tors *and for an* R-*homomorphism* $\alpha: M \to N$ *the following conditions are equivalent:*

(1) *There exists a unique* R-*homomorphism* $\delta: N \to Q_\tau(M)$ *making the diagram*

$$\begin{array}{ccc} M & \xrightarrow{\alpha} & N \\ & {\scriptstyle \hat{\tau}_M} \searrow & \downarrow{\scriptstyle \delta} \\ & & Q_\tau(M) \end{array}$$

commute, and indeed δ *is an isomorphism.*

(2) (i) $\ker(\alpha)$ <u>and</u> $\operatorname{coker}(\alpha)$ <u>are</u> τ-<u>torsion</u>;

(ii) N <u>and</u> $E(N)/N$ <u>are</u> τ-<u>torsion-free</u>.

PROOF: Let $\overline{M} = M/T_\tau(M)$.

(1) ⇒ (2): By (1), $\ker(\alpha) = \ker(\hat{\tau}_M)$, $\operatorname{coker}(\alpha) \cong E_\tau(\overline{M})/\overline{M}$, $N \cong Q_\tau(M)$, and $E(N)/N \cong E(\overline{M})/E_\tau(\overline{M})$, so the results follow from the definitions of $Q_\tau(\_)$ and $E_\tau(\_)$.

(2) ⇒ (1): Since $\ker(\alpha)$ is τ-torsion, we have $\ker(\alpha) \subseteq T_\tau(M) = \ker(\hat{\tau}_M)$. Indeed, $\ker(\alpha)$ must then equal $\ker(\hat{\tau}_M)$ or else we would have a nonzero R-homomorphism $T_\tau(M) \to N$, contradicting the fact that N is τ-torsion-free. Therefore α and $\hat{\tau}_M$ induce monomorphisms $\tilde{\alpha}\colon \overline{M} \to N$ and $\tilde{\tau}_M\colon \overline{M} \to Q_\tau(M)$, respectively. We know that $Q_\tau(M)$ is τ-injective. By (2) and Proposition 4.1, so is N. Therefore there exist R-homomorphisms δ and δ' making the diagram

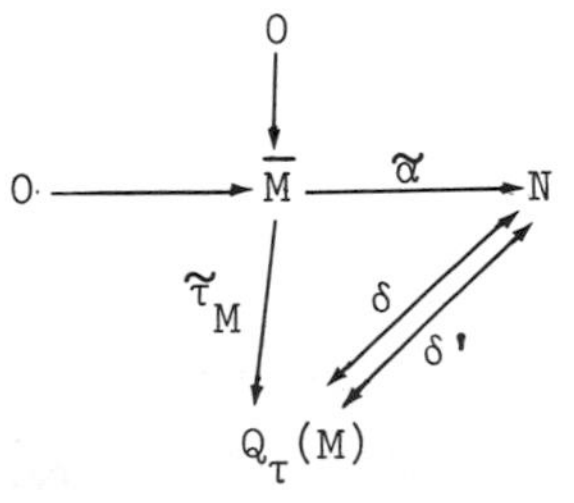

commute. Indeed, δ and δ' are unique by Proposition 5.1. Moreover, for the same reason $\delta'\delta$ must be equal to the identity map on $Q_\tau(M)$ and $\delta\delta'$ must be equal to the identity map on N. Therefore δ is an isomorphism. □

(6.4) PROPOSITION: *Let* $\tau \in$ R-tors. *For each left* R-*module* M, *the functions* $N \mapsto Q_\tau(N)$ *and* $Y \mapsto Y\hat{\tau}_M^{-1}$ *give a bijective correspondence between the sets of* τ-*pure submodules of* M *and* $Q_\tau(M)$.

PROOF: If N is a τ-pure submodule of M and if $\nu: M \to M/N$ is the canonical surjection, then $T_\tau(M)\nu \subseteq M/N \in \mathcal{F}_\tau$ and so $T_\tau(M) \subseteq N$. Therefore, $N/T_\tau(M)$ is τ-pure in $M/T_\tau(M)$ if and only if N is τ-pure in M, and so without loss of generality we can reduce to the case $T_\tau(M) = 0$. Thus $\hat{\tau}_M: M \to Q_\tau(M)$ is a monomorphism. Identifying M with its image under $\hat{\tau}_M$, we see that for every submodule Y of $Q_\tau(M)$, we have $Y\hat{\tau}_M^{-1} = Y \cap M$.

If N is a τ-pure submodule of M, then by Proposition 6.1 we have an exact sequence $0 \to Q_\tau(N) \to Q_\tau(M) \to Q_\tau(M/N)$ and so $Q_\tau(M)/Q_\tau(N)$ is isomorphic to a submodule of $Q_\tau(M/N)$ and thus is τ-torsion-free. Therefore $Q_\tau(N)$ is τ-pure in $Q_\tau(M)$.

If Y is a τ-pure submodule of $Q_\tau(M)$ then $M/[Y \cap M] \cong [M + Y]/Y$, which is a submodule of the τ-torsion-free left R-module $Q_\tau(M)/Y$. Therefore $Y \cap M = Y\hat{\tau}_M^{-1}$ is τ-pure in M.

Again, if N is a τ-pure submodule of M then $[M \cap Q_\tau(N)]/N$ is a submodule of M/N and so is τ-torsion-free. On the other hand, it is also a submodule of $Q_\tau(N)/N$ and so is τ-torsion. Therefore, we must have $N = M \cap Q_\tau(N)$ and so the map $N \mapsto Q_\tau(N) \mapsto M \cap Q_\tau(N)$ is the identity.

Finally, if Y is a τ-pure submodule of $Q_\tau(M)$ then $Y/[Y \cap M] \cong [Y + M]/M$, which is a submodule of the τ-torsion module $Q_\tau(M)/M$. Thus $Y/[Y \cap M] \subseteq Q_\tau(Y \cap M)/[Y \cap M]$. If $\lambda\colon Y \to Q_\tau(Y \cap M)$ is the inclusion map then the diagram

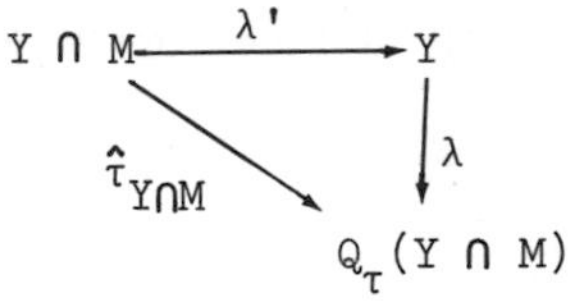

commutes, where λ' is also the inclusion map. Since Y is a submodule of $Q_\tau(M)$, it is τ-torsion-free. Moreover, by the exactness of the sequence
$0 \to Q_\tau(M)/Y \to E(M)/Y \to E(M)/Q_\tau(M) \to 0$ and the fact that $Q_\tau(M)/Y$ is τ-torsion-free by the τ-purity of Y and that $E(M)/Q_\tau(M)$ is τ-torsion-free by the absolute τ-purity of $Q_\tau(M)$, we have that $E(M)/Y \in \mathcal{F}_\tau$. Then $E(Y)/Y \in \mathcal{F}_\tau$ since it is a submodule of $E(M)/Y$. By Proposition 6.3 the R-homomorphism λ is therefore an isomorphism, and so $Y = Q_\tau(Y \cap M)$. Therefore, the map $Y \mapsto Y \cap M \mapsto Q_\tau(Y \cap M)$ is the identity, so the correspondence is bijective. □

We have so far associated with each left R-module M a localization module $Q_\tau(M)$. In particular, considering R

as a left R-module over itself, $Q_\tau(R)$ is a left R-module. We should in fact like to make $Q_\tau(R)$ into a ring.

(6.5) PROPOSITION: <u>Let</u> $\tau \in$ R-tors <u>and let</u> R_τ <u>be the endomorphism ring of the left</u> R-<u>module</u> $Q_\tau(R)$. <u>Then</u> R_τ <u>is canonically a left</u> R-<u>module which is isomorphic to</u> $Q_\tau(R)$.

PROOF: If $x \in Q_\tau(R)$ then x induces an R-homomorphism $\rho_x\colon R \to Q_\tau(R)$ defined by $r \mapsto rx$. Since $Q_\tau(R)$ is τ-torsion-free, $T_\tau(R) \subseteq \ker(\rho_x)$ and so ρ_x induces an R-homomorphism $\bar{\rho}_x\colon R/T_\tau(R) \to Q_\tau(R)$. By Proposition 5.1, $\bar{\rho}_x$ can be uniquely extended to an R-homomorphism $\beta_x\colon Q_\tau(R) \to Q_\tau(R)$.

If $r \in R$, let $\bar{r} = r + T_\tau(R)$. We can then define a left R-module structure on R_τ by setting $r\alpha = \beta_{\bar{r}}\alpha$ for every $r \in R$ and $\alpha \in R_\tau$. The verification that this does in fact make R_τ into a left R-module is straightforward using the uniqueness property in Proposition 5.1.

Now let $\theta\colon Q_\tau(R) \to R_\tau$ be the function defined by $x \mapsto \beta_x$. We claim that θ is an R-homomorphism. Indeed, if $x, y \in Q_\tau(R)$ then $\beta_x + \beta_y$ and β_{x+y} both extend $\bar{\rho}_x + \bar{\rho}_y$ and so, by Proposition 5.1, must be equal. Similarly, if $r \in R$ and $x \in Q_\tau(R)$ then $r\beta_x$ and β_{rx} both extend $\bar{\rho}_{rx}$ and so, again, must be equal. Thus θ is an R-homomorphism. Furthermore, θ is in fact an

R-epimorphism, for if $\alpha \in R$ then $\alpha = \beta_{\overline{1\alpha}}$. Finally, θ is monic since $x \neq 0$ implies that $\overline{1}\beta_x = x \neq 0$ and so $\beta_x \neq 0$. Thus θ is an isomorphism. □

The isomorphism θ from Proposition 6.5 together with the R-homomorphism $\hat{\tau}_R$ give us an R-homomorphism $\hat{\tau}: R \to R_\tau$. The ring R_τ is called the localization of the ring R at τ.

(6.6) PROPOSITION: *For* $\tau \in$ R-tors,

(1) $\hat{\tau}$ *is a ring homomorphism;*

(2) *Every absolutely* τ*-pure left* R*-module has the structure of a left* R_τ*-module which naturally extends its structure as a left* R*-module;*

(3) *If* $\alpha: M \to N$ *is an* R*-homomorphism between left* R_τ*-modules such that* $R_\tau(M\alpha)$ *is* τ*-torsion-free as a left* R*-module, then is naturally an* R_τ*-homomorphism;*

(4) *Any* R*-homomorphism between absolutely* τ*-pure left* R*-modules is naturally an* R_τ*-homomorphism.*

PROOF: (1) Let $r, r' \in R$. Then $\hat{\tau}(rr')$ and $\hat{\tau}(r)\hat{\tau}(r')$ are both endomorphisms of $Q_\tau(R)$ that extend the R-homomorphism $R/T_\tau(R) \to Q_\tau(R)$, defined by $a + T_\tau(R) \mapsto arr' + T_\tau(R)$, and so, by Proposition 5.1, are equal. Thus $\hat{\tau}$ is a ring homomorphism.

(2) If $N \in \mathcal{E}_\tau$ and $x \in N$, then x defines an R-homomorphism $R \to N$ given by $r \mapsto rx$. Since N is τ-torsion-free, $T_\tau(R)x = 0$, and so this induces an R-homomorphism $R/T_\tau(R) \to N$ that is in turn uniquely extendable to an R-homomorphism $\zeta_x\colon Q_\tau(R) \to N$. If $\alpha \in R_\tau$, define $\alpha \cdot x$ to be $(1)\alpha\zeta_x \in N$. This makes N a left R_τ-module. Furthermore, if $\alpha = \hat{\tau}(r)$ for $r \in R$, then $\alpha \cdot x = rx$ and so the R_τ-module structure naturally extends the R-module structure of N.

(3) For any $m \in M$, define the map $\alpha_m\colon R_\tau/\mathrm{im}(\hat{\tau}) \to R_\tau(M\alpha)$ by $\alpha_m\colon q + \mathrm{im}(\hat{\tau}) \mapsto (qm)\alpha - q(m\alpha)$. Then α_m is an R-homomorphism. But by construction $R_\tau/\mathrm{im}(\hat{\tau})$ is τ-torsion and $R_\tau(M\alpha)$ is τ-torsion-free, so α_m must be the zero map. This proves that α is an R_τ-homomorphism.

(4) This is just a special case of (3). □

Since we will often be called upon to make the passage from R_τ-modules to R-modules and back, it is worth digressing a bit here to check when the important homological properties--injectivity and projectivity--are preserved under such transitions.

(6.7) PROPOSITION: Let $\tau \in$ R-tors and let M be

a left R_τ*-module that is* τ*-torsion-free as a left* R*-module. Then the following conditions are equivalent:*

(1) M *is injective as a left* R*-module*.

(2) M *is injective as a left* R_τ*-module*.

PROOF: (1) ⇒ (2): Let E be the injective hull of M in R_τ-mod. Since M is injective as a left R-module, there exists an R-homomorphism $\alpha: E \to M$ which restricted to M is the identity map. By Proposition 6.6(3), α is then an R_τ-homomorphism and so M is a direct summand of E in R_τ-mod. Thus we must have M = E, proving (2).

(2) ⇒ (1): Let E be the injective hull of M in R-mod. Then M is τ-torsion-free as a left R-module and hence so is E. Therefore, E is absolutely τ-pure. This implies that $E = Q_\tau(E)$. By Proposition 6.6(3), the inclusion map $M \to E$ is an R -homomorphism. Therefore by (2) there exists an R_τ-homomorphism $\alpha: E \to M$ the restriction of which to M is the identity map. Thus M is a direct summand of E in R-mod and so M = E, proving (1). □

(6.8) COROLLARY: *Let* $\tau \in$ R-tors *and let* M *be a left* R_τ*-module which is* τ*-torsion-free as a left* R*-module. Then the injective hull of* ${}_RM$ *equals the injective hull of* ${}_{R_\tau}M$.

PROOF: Since M is τ-torsion-free as a left R-module, $E({}_R M)$ is absolutely τ-pure and so by Proposition 6.7 is an injective left R_τ-module. Since M is large in $E({}_R M)$, this must in fact be the injective hull of ${}_{R_\tau}M$. □

(6.9) PROPOSITION: Let $\tau \in R\text{-tors}$ and let M be a left R_τ-module that is projective as a left R-module. Then M is projective as a left R_τ-module.

PROOF: Since M is projective as a left R-module, by the Dual Basis Theorem there exists a subset $\{m_i\}$ of M and a subset $\{\alpha_i\}$ of $\mathrm{Hom}_R(M,R)$, where the indices range over some index set Ω, such that for every $m \in M$, we have $m\alpha_i = 0$ for all but finitely many indices $i \in \Omega$ and $m = \Sigma(m\alpha_i)m_i$. If $r \in R$ then $rm = \hat{\tau}(r)m$ for all $m \in M$ and so $m = \Sigma\hat{\tau}(m\alpha_i)m_i$. By Proposition 6.6, the map $m \mapsto \hat{\tau}(m\alpha_i)$ is an R_τ-homomorphism for each $i \in \Omega$ and so, by the Dual Basis Theorem, M is projective as a left R_τ-module. □

Let E be an injective left R-module with endomorphism ring S and let $B = \mathrm{Hom}_S(E,E)$. Then $E \in B\text{-mod-}S$. Moreover, we have a canonical ring homomorphism $\delta_E\colon R \to B$ defined by $r \mapsto \lambda_r$ where $\lambda_r x = rx$. The ring B is called the <u>bicommutator</u> of E.

(6.10) PROPOSITION: <u>If</u> $E \in \tau \in R\text{-tors}$ <u>and</u> B <u>is the bicommutator of</u> E, <u>then</u> B <u>is absolutely</u> τ-<u>pure as a left</u> R-<u>module</u>.

PROOF: If $0 \neq \beta \in B$ then there exists an $x \in E$ with $\beta x \neq 0$. Consider the R-homomorphism $\varphi: B \to E$ defined by $\beta' \mapsto \beta' x$. Then $\beta \notin \ker(\varphi)$ and so $\beta \notin T_\tau({}_RB)$. Since this is true for any β, we must have ${}_RB \in \mathcal{F}_\tau$.

Let $I \in \mathcal{L}_\tau$ and let $\alpha: I \to B$ be an R-homomorphism. For any $x \in E$ consider the R-homomorphism $\psi_x: I \to E$ defined by $\psi_x: a \mapsto (a\alpha)x$. Since E is injective, ψ_x can be extended to an R-homomorphism $R \to E$, and so there exists an $x' \in E$ with $\psi_x: a \mapsto ax'$. Indeed, it is straightforward to check that the map $\beta: x \mapsto x'$ is an endomorphism of ${}_SE$, i.e., an element of B. Therefore $(a\alpha)x = ax' = a(\beta x) = (a\beta)x$ for all $x \in E$ and so $\alpha: a \mapsto a\beta$, which is clearly extendable to an R-homomorphism $R \to E$. Thus B is τ-injective. □

(6.11) PROPOSITION: <u>If</u> $E \in \tau \in R\text{-tors}$ <u>then</u> $\ker(\delta_E) = T_\tau(R)$.

PROOF: Since $\ker(\delta_E) = \{r \in R \mid rE = 0\}$ we immediately have $\ker(\delta_E) \supseteq T_\tau(R)$. On the other hand, if $\alpha \in \mathrm{Hom}_R(\ker(\delta_E), E)$ then α can be extended to an R-homomorphism $R \to E$, and so there exists an $x \to E$ with $a\alpha = ax$ for all $a \in \ker(\delta_E)$. But $ax = 0$ for all such a

by definition, and so $\alpha = 0$, proving that $\ker(\delta_E)$ is τ-torsion. Thus we have the reverse containment and so equality. □

(6.12) PROPOSITION: *Let* $E \in \tau \in R$-tors *and let* B *be the bicommutator of* E. *Then there exists a unique ring homomorphism* γ *making the diagram*

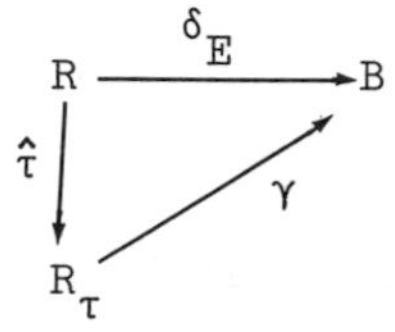

commute. Moreover, γ *is a monomorphism*.

PROOF: Let S be the endomorphism ring of E. Since E is absolutely τ-pure, it is canonically a left R_τ-module. If $\alpha \in S$ and $x \in E$, then we have an R-homomorphism $\mathrm{im}(\hat{\tau}) \to E$ defined by $r + T(R) \mapsto r(x\alpha)$. The R-homomorphisms $\beta, \beta' \colon R_\tau \to E$, defined respectively by $q \mapsto q(x\alpha)$ and $q \mapsto (qx)\alpha$, both extend this homomorphism and so, by Proposition 5.1, they must be equal. This shows that $E \in R_\tau$-mod-S.

Now define $\gamma\colon R_\tau \to B$ by $\gamma(q)\colon x \mapsto qx$. Then γ is a ring homomorphism and the above diagram commutes. Moreover, if $\gamma'\colon R_\tau \to B$ is another ring homomorphism making the diagram commute, then $\gamma'(rq) = \gamma'(\hat{\tau}(r)q) = \gamma'\hat{\tau}(r) \cdot \gamma'(q) = \delta_E(r)\gamma'(q) = r\gamma'(q)$ for all $r \in R$ and $q \in R_\tau$, and so γ'

is an R-homomorphism $R_\tau \to B$. Furthermore, γ and γ' agree on $\mathrm{im}(\hat{\tau})$ and so, by Proposition 5.1, they must be equal.

Finally, to see that γ is a monomorphism we note that $\mathrm{im}(\hat{\tau})$ is large in R_τ and that
$\ker(\gamma) \cap \mathrm{im}(\hat{\tau}) = \{\hat{\tau}(r) \mid \delta_E(r) = 0\} = \hat{\tau}(T_\tau(R)) = 0$. □

(6.13) PROPOSITION: <u>If</u> $x \in E \in \tau \in$ R-tors <u>and if</u> B <u>is the bicommutator of</u> E <u>then</u> Rx <u>is</u> τ-<u>dense in</u> Bx.

PROOF: Let α: Bx/Rx $\to$ E be an R-homomorphism and let ν: Bx $\to$ Bx/Rx be the canonical surjection. Then $\nu\alpha$ can be extended to an R-endomorphism α' of E, and for any $\beta \in B$, we have $(\beta x)\nu\alpha = \beta(x\alpha') = \beta(x\nu\alpha) = 0$. Thus $\alpha = 0$ and so Bx/Rx is τ-torsion. □

(6.14) PROPOSITION: <u>If</u> $E \in \tau \in$ R-tors <u>and if</u> E <u>has endomorphism ring</u> S <u>and bicommutator</u> B, <u>then a sufficient condition for</u> R_τ <u>to be isomorphic to</u> B <u>is that</u> E <u>be cyclic as a right</u> S-<u>module</u>.

PROOF: Assume that E = xS for some $x \in E$. Then for any $\beta \in B$, the condition $\beta x = 0$ implies that $\beta E = 0$ whence $\beta = 0$. In particular, the R-homomorphism $R_\tau \to E$ defined by $q \mapsto qx$ is an isomorphism. This shows that $R_\tau x$

is absolutely τ-pure. Since Bx is τ-torsion-free, it follows that $R_\tau x$ is τ-pure in Bx. But $Bx/R_\tau x$ is a homomorphic image of Bx/Rx and so is τ-torsion by Proposition 6.13. Therefore we must have $Bx = R_\tau x$. Thus for every $\beta \in B$ there exists a $q \in R_\tau$ with $(\beta - \gamma(q))x = 0$. But, as we have seen, this implies that $\beta - \gamma(q) = 0$, and so γ is a surjection and therefore is a ring isomorphism. □

(6.15) COROLLARY: _If_ $\tau \in R$-tors _then there exists an_ $E \in \tau$ _the bicommutator of which is isomorphic to_ R_τ.

PROOF: By Proposition 6.14 it suffices to find an $E \in \tau$ that is cyclic over its endomorphism ring. Indeed, if E is any member of τ and if Ω is an index set the cardinality of which equals card(E), then E^Ω has this property. □

References for Section 6

Cunningham [32]; Cunningham, Rutter, and Turnidge [34]; Findlay and Lambek [47, 48]; Gabriel [52]; Goldman [63]; Hacque [67]; Helzer [71]; Lambek [90, 91, 92, 93]; Morita [105, 108, 109]; Sanderson [135]; Sandomierski [136]; Stenström [147]; Walker and Walker [164]; Winton [165].

CHAPTER II

THE SPACE R-tors

7. ORDER AMONG TORSION THEORIES

We can partially order the set R-tors (for purists: it is indeed a set since we have shown that it bijectively corresponds to the set of all idempotent filters of left ideals of R) by setting $\tau \geq \tau'$ if and only if $E \geq E'$ for some (and hence every) $E \in \tau$ and $E' \in \tau'$.

(7.1) PROPOSITION: The following conditions are equivalent for $\tau, \tau' \in$ R-tors:

(1) $\tau \geq \tau'$.

(2) $\mathcal{J}_\tau \supseteq \mathcal{J}_{\tau'}$.

(3) $\mathcal{L}_\tau \supseteq \mathcal{L}_{\tau'}$.

(4) $\mathcal{F}_\tau \subseteq \mathcal{F}_{\tau'}$.

(5) For every left R-module M, the identity map on M is a continuous function $X_\tau(M) \to X_{\tau'}(M)$.

PROOF: (1) ⇒ (2): Assume that $\tau \geq \tau'$ and let $M \in \mathcal{J}_{\tau'}$. If there exists an $N \in \mathcal{F}_\tau$ for which

$\mathrm{Hom}_R(M,E(N)) \neq 0$ then $E(N) \geq E$ for some $E \in \tau$ whence $\mathrm{Hom}_R(M,E) \neq 0$. Since $\tau \geq \tau'$ there exists an $E' \in \tau'$ with $\mathrm{Hom}_R(M,E') = 0$. But E' is τ'-torsion-free and so we have contradicted Proposition 1.3. Therefore $\mathrm{Hom}_R(M,E(N)) = 0$ for all $N \in \mathcal{F}_\tau$, which implies that M is τ-torsion. Thus we have (2).

(2) ⇔ (3): This proof is trivial.

(2) ⇒ (4): Let $N \in \mathcal{F}_\tau$. Then $\mathrm{Hom}_R(M,E(N)) = 0$ for all $M \in \mathcal{T}_\tau$. By (2), $\mathrm{Hom}_R(M,E(N)) = 0$ for all $M \in \mathcal{T}_{\tau'}$, proving that $N \in \mathcal{F}_{\tau'}$.

(4) ⇒ (1): If $E \in \tau$ then E is τ-torsion-free, and so $E \in \mathcal{F}_{\tau'}$ by (4). Therefore $E \geq E'$ for some $E' \in \tau'$.

(2) ⇔ (5): Clearly (5) is equivalent to the condition that for every submodule N of a left R-module M, $M/N \in \mathcal{T}_{\tau'}$ implies that $M/N \in \mathcal{T}_{\tau'}$ which is equivalent to (2). □

(7.2) PROPOSITION: <u>Let</u> $\tau, \tau' \in$ R-tors. <u>Then each of the following conditions implies the next. They are equivalent if</u> $T_\tau(R) \in \mathcal{T}_{\tau'}$.

(1) $\tau \leq \tau'$.

(2) <u>A left</u> R-<u>module</u> M <u>is</u> τ-<u>injective if and only if</u> $\mathrm{Ext}^1_R(R/H,M) = 0$ <u>for all</u> $H \in \mathcal{L}_\tau \cap \mathcal{L}_{\tau'}$.

(3) $Q_\tau(R)/K \notin \mathcal{F}_{\tau'}$, <u>for every</u> ${}_RK \subset Q_\tau(R)$ <u>with</u> $K\hat{\tau}_R^{-1} \in \mathcal{L}_\tau$.

PROOF: (1) ⇒ (2): Suppose that $\mathrm{Ext}_R^1(R/H,M) = 0$ for every left ideal $H \in \mathcal{L}_\tau \cap \mathcal{L}_{\tau'}$. Let I be a proper left ideal of R belonging to $\mathcal{L}_\tau$ and let $\alpha \in \mathrm{Hom}_R(I,M)$. Then (1) implies that $R/I \in \mathcal{T}_\tau \subseteq \mathcal{T}_{\tau'}$ and so I belongs to $\mathcal{L}_\tau \cap \mathcal{L}_{\tau'}$, whence $\mathrm{Ext}_R^1(R/I,M) = 0$. By the exactness of the sequence $\mathrm{Hom}_R(R,M) \to \mathrm{Hom}_R(I,M) \to \mathrm{Ext}_R^1(R/I,M) = 0$, this means that M is τ-injective by Proposition 4.1. The converse is trivial.

(2) ⇒ (3): Let K be a proper submodule of $Q_\tau(R)$ with $K\hat{\tau}_R^{-1}$ τ-dense in R, and assume that $Q_\tau(R)/K \in \mathcal{F}_{\tau'}$. We first claim that $\mathrm{Ext}_R^1(R/H,K) = 0$ for any left ideal $H \in \mathcal{L}_\tau \cap \mathcal{L}_{\tau'}$. Indeed, if H is such an ideal and if $\alpha \in \mathrm{Hom}_R(H,K)$, then by the τ-injectivity of $Q_\tau(R)$ there exists an R-homomorphism β making the diagram

$$\begin{array}{ccccc} 0 & \longrightarrow & H & \longrightarrow & R \\ & & \downarrow{\scriptstyle\alpha} & & \downarrow{\scriptstyle\beta} \\ 0 & \longrightarrow & K & \longrightarrow & Q_\tau(R) \end{array}$$

commute. Then β induces an R-homomorphism $\beta': R/H \to Q_\tau(R)/K$ which must be the zero map since $R/H \in \mathcal{T}_{\tau'}$ and $Q_\tau(R)/K \in \mathcal{F}_{\tau'}$. Therefore $R\beta \subseteq K$. From this we deduce that the sequence $\mathrm{Hom}_R(R,K) \to \mathrm{Hom}_R(H,K) \to 0$

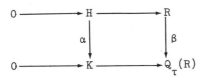

is exact. But the sequence $\mathrm{Hom}_R(R,K) \to \mathrm{Hom}_R(H,K) \to \mathrm{Ext}^1_R(R/H,K) \to \mathrm{Ext}^1_R(R,K) = 0$ is also exact and so we must have $\mathrm{Ext}^1_R(R/H,K) = 0$. By (2), this implies that ${}_RK$ is τ-injective and hence absolutely τ-pure. In particular, K is τ-pure in $K + R\hat{\tau}_R$. On the other hand, $[K + R\hat{\tau}_R]/K \cong R\hat{\tau}_R/[K \cap R\hat{\tau}_R]$ which is a homomorphic image of $R/K\hat{\tau}_R^{-1} \in \mathcal{J}_\tau$ and so K is τ-dense in $K + R\hat{\tau}_R$. Therefore $K = K + R\hat{\tau}_R$ and so $R\hat{\tau}_R \subseteq K$. Thus $Q_\tau(R)/K$ is a nonzero homomorphic image of $Q_\tau(R)/R\hat{\tau}_R$ and so is τ-torsion, contradicting the assumption that K is τ-pure in $Q_\tau(R)$. This proves (3).

(3) $\Rightarrow$ (1): Now assume that $T_\tau(R) \in \mathcal{J}_{\tau'}$. Let $I \in \mathcal{L}_\tau$, $I \neq R$. By (3), $Q_\tau(R)/I\hat{\tau}_R$ is not τ'-torsion-free and so we have a submodule W of $Q_\tau(R)$ properly containing $I\hat{\tau}_R$ such that W is the τ'-purification of $I\hat{\tau}_R$ in $Q_\tau(R)$. Then $I \subseteq W\hat{\tau}_R^{-1}$ so $W\hat{\tau}_R^{-1} \in \mathcal{L}_\tau$. But $Q_\tau(R)/W \in \mathcal{F}_{\tau'}$, and so by (3) we must have $W = Q_\tau(R)$. Therefore, $I\hat{\tau}_R$ is τ'-dense in $Q_\tau(R)$ and so is τ'-dense in $R\hat{\tau}_R$. Since $I\hat{\tau}_R = [I + T_\tau(R)]/T_\tau(R)$, this implies that $I + T_\tau(R)$ is τ'-dense in R. We then have the exact sequence $0 \to [I + T_\tau(R)]/I \to R/I \to R/[I + T_\tau(R)] \to 0$ where $[I + T_\tau(R)]/I \cong T_\tau(R)/[I \cap T_\tau(R)]$ is τ'-torsion by our additional hypothesis. Therefore $R/I \in \mathcal{J}_{\tau'}$. We have thus shown that $\mathcal{L}_\tau \subseteq \mathcal{L}_{\tau'}$, proving that $\tau \leq \tau'$ by Proposition 7.1. □

(7.3) PROPOSITION: If $\tau \leq \tau'$ in R-tors, then for every $M \in$ R-mod there exists a unique R-homomorphism $\alpha_M\colon Q_\tau(M) \to Q_{\tau'}(M)$ making the diagram

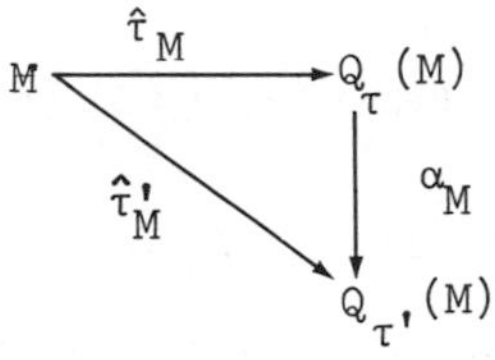

commute. Furthermore, $\ker(\alpha_M) = Q_\tau(T_{\tau'}(M)/T_\tau(M))$.

PROOF: Let $\overline{M} = M/T_\tau(M)$. Then τ_M induces an R-homomorphism $\overline{\tau}\colon \overline{M} \to Q_\tau(M)$. Moreover, $Q_\tau(M)/\overline{M\tau} = Q_\tau(M)/M\hat{\tau}_M \in \mathcal{J}_\tau \subseteq \mathcal{J}_{\tau'}$, therefore, by the τ'-injectivity of $Q_{\tau'}(M)$ there exists at least one R-homomorphism α_M that makes the diagram commute, and indeed it is unique by Proposition 5.1.

Moreover, if $\overline{\overline{M}} = M/T_{\tau'}(M)$ we have the exact sequence $0 \to T_{\tau'}(M)/T_\tau(M) \to \overline{M} \to \overline{\overline{M}} \to 0$ and, applying $Q_\tau(\_)$, we obtain the exact sequence $0 \to Q_\tau(T_{\tau'}(M)/T_\tau(M)) \to Q_\tau(\overline{M}) \xrightarrow{\pi} Q_\tau(\overline{\overline{M}})$. Since $\tau \leq \tau'$, we have $\overline{\overline{M}} \in \mathcal{F}_{\tau'} \subseteq \mathcal{F}_\tau$ and so $\overline{\overline{M}}$ is a large submodule of $Q_\tau(\overline{\overline{M}})$. Therefore, $Q_\tau(\overline{\overline{M}}) \in \mathcal{F}_{\tau'}$ and so the R-homomorphism $\alpha_{\overline{\overline{M}}}\colon Q_\tau(\overline{\overline{M}}) \to Q_{\tau'}(\overline{\overline{M}})$ is a monomorphism.

By uniqueness, we then have the commutative diagram

$$\begin{array}{ccccc} Q_\tau(\bar{M}) & \xrightarrow{\pi} & Q_\tau(\bar{\bar{M}}) & \xrightarrow{\alpha_{\bar{\bar{M}}}} & Q_{\tau'}(\bar{\bar{M}}) \\ \downarrow{=} & & & & \downarrow{=} \\ Q_\tau(M) & & \xrightarrow{\alpha_M} & & Q_{\tau'}(M) \end{array}$$

and so $\ker(\alpha_M) = \ker(\pi\alpha_{\bar{\bar{M}}}) = \ker(\pi) = Q_\tau(T_{\tau'}(M)/T_\tau(M))$. □

If $\tau \leq \tau'$ we say that τ' is a <u>generalization</u> of and that τ is a <u>specialization</u> of τ'. We will denote the set of all generalizations of τ by $\text{gen}(\tau)$ and all specializations of τ' by $\text{spcl}(\tau')$.

(7.4) PROPOSITION: <u>The following conditions are equivalent for</u> $\tau \in R\text{-tors}$:

(1) τ <u>is a generalization of the Goldie torsion theory</u> τ_G (see Chapter VII, Example 7).

(2) <u>Every left</u> R-<u>module is</u> τ-<u>dense in its injective hull</u>.

(3) <u>Every absolutely</u> τ-<u>pure left</u> R-<u>module is injective</u>.

PROOF: (1) ⇒ (2): Let $E \in \tau$. Since $\tau \geq \tau_G$, E is τ_G-torsion-free and so is nonsingular. Now let M be a left R-module and let $\alpha \in \text{Hom}_R(E(M)/M, E)$. If $x \in E(M)$ with $(x + M)\alpha \neq 0$, then $(0:(x + M)\alpha) \supseteq (M:x)$ which is large in R since M is large in $E(M)$. Therefore, we have a contradiction and so $\alpha = 0$. This proves that M is

τ-dense in $E(M)$.

(2) $\Rightarrow$ (3): If M is an absolutely pure left R-module, then $E(M)/M$ is τ-torsion by (2) and τ-torsion-free by definition, so $E(M)/M = 0$, proving the injectivity of M.

(3) $\Rightarrow$ (1): Let $E \in \tau$ and let I be a large left ideal of R. Then $R/I \subseteq E(I)/I \in \mathcal{J}_\tau$ by (3), so $\mathrm{Hom}_R(R/I,E) = 0$. This implies that $Ix \neq 0$ for every $x \in E$, whence E is nonsingular and therefore τ_G-torsion-free. Thus $\mathcal{F}_\tau \subseteq \mathcal{F}_{\tau_G}$ and so τ is a generalization of τ_G. □

It is clear that under this partial ordering R-tors has a unique minimal member ξ and a unique maximal member χ, defined respectively by $\mathcal{J}_\xi = \{0\}$ and $\mathcal{J}_\chi$ = R-mod. That is to say, ξ consists of all injective cogenerators of R-mod and χ consists only of the 0 module.

A torsion theory $\tau \in$ R-tors is said to be <u>nontrivial</u> if and only if $\tau \neq \xi$; it is said to be <u>proper</u> if and only if $\tau \neq \chi$. We shall denote the set of all proper torsion theories on R-mod by R-prop.

<u>References for Section 7</u>

Dickson [36]; Gabriel [52]; Golan [55]; Golan and Teply [57]; Goldman [63]; Hacque [67]; Jans [79]; Lambek [92]; Raynaud [125]; Stenström [147].

8. LATTICE STRUCTURE ON R-tors

We now want to canonically define on R-tors the structure of a complete lattice compatible with the partial order defined in the previous section. If U is a subset of R-tors and if $E_\tau \in \tau$ for every $\tau \in U$, let $E = \Pi\{E_\tau \mid \tau \in U\}$. We call the equivalence class of E the <u>meet</u> of U and denote it by $\wedge U$.

(8.1) PROPOSITION: <u>If</u> $U \subseteq$ R-tors <u>then</u>

(1) $\mathcal{J}_{\wedge U} = \cap\{\mathcal{J}_\tau \mid \tau \in U\}$;

(2) <u>If</u> $\tau' \leq \tau$ <u>for all</u> $\tau \in U$, <u>then</u> $\tau' \leq \wedge U$.

PROOF: (1) A left R-module M belongs to $\mathcal{J}_{\wedge U}$ if and only if $\mathrm{Hom}_R(M,E) = 0$, which is true if and only if $\mathrm{Hom}_R(M,E_\tau) = 0$ for all $\tau \in U$. But this is precisely the condition for M to belong to $\cap\{\mathcal{J}_\tau \mid \tau \in U\}$.

(2) Let $\tau' \leq \tau$ for all $\tau \in U$. Then $\mathcal{J}_{\tau'} \subseteq \mathcal{J}_\tau$ for all such τ and so $\mathcal{J}_{\tau'} \subseteq \cap \mathcal{J}_\tau = \mathcal{J}_{\wedge U}$, proving that $\tau' \leq \wedge U$. □

Likewise, if U is a subset of R-tors, let $V = \cap\{\mathrm{gen}(\tau) \mid \tau \in U\}$. Pick $E_{\tau'} \in \tau'$ for all $\tau' \in V$ and set $E = \Pi\{E_{\tau'} \mid \tau' \in V\}$. We call the equivalence class of E the <u>join</u> of U and denote it by $\vee U$.

(8.2) PROPOSITION: <u>If</u> $U \subseteq$ R-tors <u>then</u>

(1) $\mathcal{F}_{\vee U} = \cap\{\mathcal{F}_\tau \mid \tau \in U\}$,

(2) <u>If</u> $\tau' \geq \tau$ <u>for all</u> $\tau \in U$, <u>then</u> $\tau' \geq \vee U$.

PROOF: (1) Let $M \in \mathcal{F}_{\vee U}$. Then $E(M) \geq E$ so if τ' is the equivalence class of $E(M)$ then $\tau' \in \text{gen}(\vee U)$. Therefore $\tau' \in \text{gen}(\tau)$ for all $\tau \in U$ and so M is τ-torsion-free for all $\tau \in U$. Conversely, assume that $M \in \cap\{\mathcal{F}_\tau \mid \tau \in U\}$. If τ' is the equivalence class of $E(M)$ we then have $\tau' \in \text{gen}(\tau)$ for all $\tau \in U$ and so $\tau' \in V$. Then $E(M)$ is embeddable in a direct product of copies of E, and so $M \in \mathcal{F}_{\vee U}$.

(2) If $\tau' \geq \tau$ for all $\tau \in U$, then $\tau' \in V$. If $E' \geq \tau'$ then E' is embeddable in a direct product of copies of E, and so $E' \geq E$, proving that $\tau' \geq \vee U$. □

It is easy to see that R-tors, therefore, forms a complete lattice under $\wedge$ and $\vee$.

(8.3) PROPOSITION: <u>For</u> $\tau, \tau' \in$ R-tors <u>and</u> $U \subseteq$ R-tors,

(1) $\tau \leq \tau'$ <u>implies that</u> $\text{gen}(\tau) \supseteq \text{gen}(\tau')$;

(2) $\text{gen}(\wedge U) \supseteq \cup\{\text{gen}(\tau'') \mid \tau'' \in U\}$;

(3) $\text{gen}(\vee U) = \cap\{\text{gen}(\tau'') \mid \tau'' \in U\}$.

PROOF: (1) follows directly from the definition. By (1), we have $\text{gen}(\tau'') \subseteq \text{gen}(\wedge U)$ for every $\tau'' \in U$, from which we obtain (2). As for (3), if

$\tau_0 \in \cap\{\text{gen}(\tau'') \mid \tau'' \in U\}$, then $\tau_0 \geq \tau''$ for each $\tau'' \in U$ and so $\tau_0 \geq \vee U$, which is to say that $\tau_0 \in \text{gen}(\vee U)$. The converse is trivial. □

(8.4) PROPOSITION: _If_ $\{U_i\}$ _is a family of subsets of_ R-tors, _then_ $\wedge(\bigcup_i U_i) = \bigwedge_i(\wedge U_i)$.

PROOF: Let $U = \bigcup_i U_i$. If $\tau \leq \tau'$ for every $\tau' \in U$, then $\tau \leq \tau''$ for every $\tau'' \in U_i$, and so $\tau \leq \wedge U_i$. Since this is true for each i, then $\tau \leq \bigwedge_i(\wedge U_i)$. Hence in particular $\wedge U \leq \bigwedge_i(\wedge U_i)$. Conversely, if $\tau \leq \wedge U_i$ for each i and if $\tau' \in U$, then $\tau' \in U_i$ for some i and so $\tau \leq \wedge U_i \leq \tau'$. Thus $\tau \leq \wedge U$. In particular $\bigwedge_i(\wedge U_i) \leq \wedge U$ and so we have equality. □

For any family $\mathcal{A}$ of left R-modules, we defined in Section 1 the torsion theories $\chi(\mathcal{A})$ and $\xi(\mathcal{A})$ cogenerated and generated, respectively, by $\mathcal{A}$. We now give equivalent definitions of them in terms of the lattice structure on R-tors.

(8.5) PROPOSITION: _For any collection_ $\mathcal{A}$ _of left_ R-_modules_:

(1) $\chi(\mathcal{A}) = \vee\{\tau \in \text{R-tors} \mid \mathcal{A} \subseteq \mathcal{F}_\tau\}$;

(2) $\xi(\mathcal{A}) = \wedge\{\tau \in \text{R-tors} \mid \mathcal{A} \subseteq \mathcal{T}_\tau\}$.

PROOF: (1) We have seen that $\mathcal{A} \subseteq \mathcal{F}_{\chi(\mathcal{A})}$ for any collection $\mathcal{A}$ of left R-modules. Furthermore, if

$\mathcal{A} \subseteq \mathcal{F}_\tau$ for some $\tau \in$ R-tors and if $\mathcal{A}_0$ is a complete set of representatives of isomorphism classes of cyclic submodules of members of $\mathcal{A}$, then $\mathcal{A}_0 \subseteq \mathcal{F}_\tau$ and so $E_0 = \Pi\{E(M) \mid M \in \mathcal{A}_0\} \in \mathcal{F}_\tau$. Thus $\mathcal{F}_{\chi(\mathcal{A})} \subseteq \mathcal{F}_\tau$ and so $\chi(\mathcal{A}) \geq \tau$. By Proposition 8.2(2), therefore, we deduce (1).

(2) We have seen that $\mathcal{A} \subseteq \mathcal{T}_{\xi(\mathcal{A})}$. In a manner similar to the proof of Proposition 1.6, we can check that $\mathcal{T}_{\xi(\mathcal{A})} \subseteq \mathcal{T}_\tau$ for any $\tau \in$ R-tors with $\mathcal{A} \subseteq \mathcal{T}_\tau$. Then (2) follows by Proposition 8.1(2). □

In particular, $\chi = \chi(0) = \vee$(R-tors) and $\xi = \xi(0) = \wedge$(R-tors).

(8.6) PROPOSITION: <u>For</u> $M \in$ R-mod, <u>the following conditions hold</u>.

(1) <u>If</u> N <u>is a submodule of</u> M, <u>then</u> $\chi(N) \geq \chi(M) \geq \chi(N) \wedge \chi(M/N)$.

(2) <u>If</u> N <u>is a submodule of</u> M, <u>then</u> $\xi(M) = \xi(N) \vee \xi(M/N)$.

(3) <u>If</u> $\{M_i\}$ <u>is a family of submodules of</u> M <u>satisfying</u> $M = \cup M_i$ <u>or</u> $M = \Sigma M_i$, <u>then</u> $\chi(M) = \wedge\chi(M_i)$.

(4) <u>If</u> $\{M_i\}$ <u>is a family of submodules of</u> M <u>satisfying</u> $M = \cup M_i$ <u>or</u> $M = \Sigma M_i$, <u>then</u> $\xi(M) = \vee\xi(M_i)$.

(5) _If_ $\{M_i\}$ _is a family of submodules of_ M _satisfying_ $M = \Pi M_i$, _then_ $\chi(M) = \wedge\chi(M_i)$.

(6) _If_ N _is a large submodule of_ M, _then_ $\chi(N) = \chi(M)$.

PROOF: The proofs are direct consequences of the definitions and of Proposition 8.5. □

A torsion theory $\tau \in R\text{-tors}$ is said to be _basic_ if and only if it is of the form $\xi(R/I)$ for some proper left ideal I of R. Also, τ is said to be _cobasic_ if and only if it is of the form $\chi(R/I)$ for some proper left ideal I of R.

(8.7) PROPOSITION: _The set of all basic_ $\tau \in R\text{-tors}$ _is closed under finite joins._

PROOF: Let I and I' be proper left ideals of R. We have a canonical R-monomorphism $\alpha: R/[I \cap I'] \to R/I \oplus R/I'$. By Proposition 8.6(2), we then have $\xi(R/[I \cap I']) \leq \xi(R/I \oplus R/I') = \xi(R/I) \vee \xi(R/I')$. On the other hand, R/I and R/I' are both homomorphic images of $R/[I \cap I']$, so by Proposition 8.6(2) we have $\xi(R/I), \xi(R/I') \leq \xi(R/[I \cap I'])$ and so $\xi(R/I) \vee \xi(R/I') \leq \xi(R/[I \cap I'])$, proving equality. □

Torsion theories $\tau, \tau' \in R\text{-tors}$ are said to be _disjoint_ if and only if $\tau \wedge \tau' = \xi$.

(8.8) PROPOSITION: <u>The following conditions are equivalent for</u> $\tau, \tau' \in$ R-tors:

(1) τ <u>and</u> τ' <u>are disjoint</u>.

(2) <u>If</u> $I \in \mathcal{L}_\tau$ <u>and</u> $I' \in \mathcal{L}_{\tau'}$, <u>then</u> $I + I' = R$.

(3) $\mathcal{J}_\tau \cap \mathcal{J}_{\tau'}$ <u>does not contain a simple module</u>.

PROOF: (1) ⇒ (2): Let $I \in \mathcal{L}_\tau$ and $I' \in \mathcal{L}_{\tau'}$. Then $R/[I + I']$ is a homomorphic image of both R/I and R/I' and so belongs to $\mathcal{J}_\tau \cap \mathcal{J}_{\tau'} = \mathcal{J}_\xi = \{0\}$. Hence $R = I + I'$.

(2) ⇒ (3): Let M be a simple left R-module contained in $\mathcal{J}_\tau \cap \mathcal{J}_{\tau'}$. Then $M \cong R/I$ for some maximal left ideal I of R. Moreover, $I \in \mathcal{L}_\tau \cap \mathcal{L}_{\tau'}$ and so by (2) $I = I + I = R$, which is a contradiction.

(3) ⇒ (1): Let $0 \neq M \in \mathcal{J}_\tau \cap \mathcal{J}_{\tau'}$. Then any nonzero cyclic submodule of M has a nonzero simple homomorphic image that must also belong to $\mathcal{J}_\tau \cap \mathcal{J}_{\tau'}$, contradicting (3). Therefore, we must have $\tau \wedge \tau' = \xi$. □

(8.9) COROLLARY: <u>Let</u> $\tau, \tau' \in$ R-tors <u>be disjoint and let</u> $0 \to N \to M \to M/N \to 0$ <u>be an exact sequence in</u> R-mod <u>with</u> $N \in \mathcal{J}_\tau$ <u>and</u> $M/N \in \mathcal{J}_{\tau'}$. <u>Then</u> $N = T_\tau(M)$.

PROOF: Since N is τ-torsion, $N \subseteq T_\tau(M)$. For each $m \in T_\tau(M)$, we have $(N:m) \in \mathcal{L}_{\tau'}$, and so, by Proposition 8.8(2), $(N:m) = (N:m) + (0:m) = R$. Therefore,

$Rm \subseteq N$ and so $N = T_\tau(M)$. □

(8.10) PROPOSITION: *If* $\tau, \tau' \in$ R-tors *then* $\tau \vee \tau' = \xi(\{M \in \text{R-mod} \mid M/T_\tau(M)$ *is* τ'-*torsion*$\})$.

PROOF: Let $\mathcal{A} = \{M \in \text{R-mod} \mid M/T_\tau(M)$ is τ'-torsion$\}$. Trivially $\mathcal{J}_\tau \subseteq \mathcal{A}$. If $M \in \mathcal{J}_{\tau'}$ then $M/T_\tau(M) \in \mathcal{J}_{\tau'}$ by closure under taking homomorphic images and so $\mathcal{J}_{\tau'} \subseteq \mathcal{A}$. This shows that $\tau, \tau' \leq \xi(\mathcal{A})$ and hence $\tau \vee \tau' \leq \xi(\mathcal{A})$.

On the other hand, if $M \in \mathcal{A}$ then $M/T_\tau(M) \in \mathcal{J}_{\tau'} \subseteq \mathcal{J}_{\tau\vee\tau'}$ and $T_\tau(M) \in \mathcal{J}_\tau \subseteq \mathcal{J}_{\tau\vee\tau'}$ and so, by the closure of $\mathcal{J}_{\tau\vee\tau'}$ under taking extensions, we have $M \in \mathcal{J}_{\tau\vee\tau'}$. Hence $\mathcal{A} \subseteq \mathcal{J}_{\tau\vee\tau'}$ and so $\xi(\mathcal{A}) \leq \tau \vee \tau'$, proving equality. □

(8.11) PROPOSITION: *The lattice* R-tors *is distributive*.

PROOF: It suffices to show that for every τ, $\tau', \tau'' \in$ R-tors, we have $(\tau \vee \tau') \wedge \tau'' \leq \tau \vee (\tau' \wedge \tau'')$. By Proposition 7.1, we need to show that $\mathcal{F}_{\tau\vee(\tau'\wedge\tau'')} \subseteq \mathcal{F}_{(\tau\vee\tau')\wedge\tau''}$. Therefore consider $0 \neq M \in \mathcal{F}_{\tau\vee(\tau'\wedge\tau'')} = \mathcal{F}_\tau \cap \mathcal{F}_{\tau'\wedge\tau''}$ and assume that $M \notin \mathcal{F}_{(\tau\vee\tau')\wedge\tau''}$. Set $0 \neq N = T_{(\tau\vee\tau')\wedge\tau''}(M)$. Then $N \in \mathcal{J}_{\tau\vee\tau'} \cap \mathcal{J}_{\tau''}$. Moreover, since M is τ-torsion-free, so is N.

By Proposition 8.10, there exists a left R-module M' satisfying $M'/T_\tau(M') \in \mathcal{J}_{\tau'}$ and a nonzero R-homomorphism $\alpha: M' \to N$. Since N is τ-torsion-free, $T_\tau(M') \subseteq \ker(\alpha)$ and so $M'\alpha$ is a nonzero τ'-torsion submodule of N. But then $M'\alpha \in \mathcal{J}_{\tau'} \cap \mathcal{J}_{\tau''} = \mathcal{J}_{\tau' \wedge \tau''}$, contradicting the fact that $M \in \mathcal{F}_{\tau' \wedge \tau''}$. □

A subset U of R-tors is said to be <u>separated</u> if and only if, for each $\tau \in U$, the torsion theories τ and $\vee(U \setminus \{\tau\})$ are disjoint.

(8.12) PROPOSITION: <u>Let</u> $U \subseteq$ R-tors <u>be separated</u>. <u>Then</u> $\vee U = \xi(\{\oplus M_\tau \mid \tau \in U \text{ and } M_\tau \in \mathcal{J}_\tau\})$.

PROOF: First observe that since U is separated, sums of the form ΣM_τ, where the τ range over U and where $M_\tau \in \mathcal{J}_\tau$, are in fact direct.

Define the endofunctor $F(\_)$ of R-mod by $F(M) = \oplus\{T_\tau(M) \mid \tau \in U\}$ and let $\mathcal{A} = \{M \in \text{R-mod} \mid F(M) = M\}$. We first claim that $F(\_)$ is a left exact subfunctor of the identity functor on R-mod.

That it is a subfunctor of the identity follows from the fact that each $T_\tau(\_)$ is such. We now have to show that F is left exact. Let N be a submodule of a left R-module M and let $x \in N \cap F(M)$. The proof is completed if we show that $x \in F(N)$. By the definition of F, we have $x = x_1 + \ldots + x_k$ where the $x_i \in T_{\tau_i}(M)$ $(\tau_i \in U)$. We therefore need to show that $x_i \in T_{\tau_i}(N) = T_{\tau_i}(M) \cap N$ for each i. We do this by induction on k. Indeed, for $k = 1$ there is nothing to prove. Therefore, assume that whenever $y_1 + \ldots + y_{k-1} \in N$ with $y_i \in T_{\tau_i}(M)$, then each $y_i \in N$. Let $x = x_1 + \ldots + x_k \in N$ with $x_i \in T_{\tau_i}(M)$ for all i. Then there exist $I_1 \in \mathcal{L}_{\tau_1}$ and $I_k \in \mathcal{L}_{\tau_k}$ such that $I_1x_1 = 0 = I_kx_k$. Since τ_1 and τ_k are disjoint by the separation of U, then by Proposition 8.8 we have $I_1 + I_k = R$. Let $1 = a_1 + a_k$ where $a_1 \in I_1$ and $a_k \in I_k$. Then $a_k(\sum_{i=1}^{k} x_i) = \sum_{i=1}^{k-1} a_kx_i \in N$, and by the induction hypothesis, $a_kx_i \in N$ for all $1 \leq i \leq k-1$. Then $x_1 = (a_1 + a_k)x_1 = a_kx_1 \in N$, and so $x - x_1 = \sum_{i=2}^{k} x_i \in N$. Therefore, by the induction hypothesis, each $x_i \in N$ $(1 \leq i \leq k)$.

We can now apply Proposition 2.6 to get $T_{\xi(\mathcal{A})} = \bar{\bar{F}}$. Since $\mathcal{A} \subseteq \mathcal{J}_{\vee U}$, it therefore follows that $\xi(\mathcal{A}) \leq \vee U$. On the other hand, $\mathcal{J}_\tau \subseteq \mathcal{A}$ for each $\tau \in U$ and so $\tau \leq \xi(\mathcal{A})$ for each $\tau \in U$. By Proposition 8.2 this implies that $\vee\tau \leq \xi(\mathcal{A})$ and so we have equality. □

(8.13) PROPOSITION: <u>The following conditions on a subset</u> U <u>of</u> R-tors <u>are equivalent</u>:

(1) U <u>is separated</u>.

(2) <u>For any partition</u> $U = U_1 \cup U_2$ <u>of</u> U, $\vee U_1$

and vU_2 *are disjoint.*

PROOF: The proof that (2) ⇒ (1) is trivial. Conversely, let U be separated and let $U = U_1 \cup U_2$ be a partition of U. By Proposition 8.12, $vU_1 = \xi(\{\oplus T_\tau(M) \mid M \in R\text{-mod} \text{ and } \tau \in U_1\})$. For any left R-module M and any $\tau' \in U_2$, $T_{\tau'}(M) \cap [\oplus \{T_\tau(M) \mid \tau \in U_1\}] = 0$ and so $T_{\tau'}(M) \cap T_{vU_1}(M) = 0$. Since this is true for any $\tau' \in U_2$, then $[\oplus \{T_\tau(M) \mid \tau \in U_2\}] \cap T_{vU_1}(M) = 0$ whence $T_{vU_2}(M) \cap T_{vU_1}(M) = 0$. Thus vU_1 and vU_2 are disjoint. □

References for Section 8

Bronowitz [21, 22]; Dickson [36]; Golan [55, 56]; Hacque [67]; Raynaud [125]; Stenström [147].

9. CHANGE OF RINGS

Let R and S be rings and let F: R-mod → S-mod be an exact functor that commutes with direct sums. If $\sigma \in$ S-tors then it is easy to check that $\{M \in R\text{-mod} \mid F(M) \in \mathcal{T}_\sigma\}$ satisfies the conditions of Proposition 1.6(2) and so is of the form $\mathcal{T}_\tau$ for some $\tau \in$ R-tors. Thus F induces a function F^o: S-tors → R-tors.

(9.1) PROPOSITION: *Let* F: R-mod → S-mod *be an exact functor that commutes with direct sums. Then*

(1) $\sigma \leq \sigma'$ <u>in</u> S-tors <u>implies that</u> $F^{\circ}(\sigma) \leq F^{\circ}(\sigma')$ <u>in</u> R-tors;

(2) <u>If</u> $U \subseteq$ S-tors <u>then</u> $F^{\circ}(\wedge U) = \wedge F^{\circ}(U)$;

(3) <u>For any</u> $\tau \in$ R-tors,

$$(F^{\circ})^{-1}(\text{gen}(\tau)) = \text{gen}(\xi(F(\mathcal{J}_\tau))).$$

PROOF: (1) and (2) follow directly from the definitions. As for (3), $(F^{\circ})^{-1}(\text{gen}(\tau)) = \{\sigma \in \text{S-tors} \mid F^{\circ}(\sigma) \geq \tau\} = \{\sigma \in \text{S-tors} \mid M \in \mathcal{J}_\tau \Rightarrow F(M) \in \mathcal{J}_\sigma\} = \{\sigma \in \text{S-tors} \mid F(\mathcal{J}_\tau) \subseteq \mathcal{J}_\sigma\} = \{\sigma \in \text{S-tors} \mid \sigma \geq \xi(F(\mathcal{J}_\tau))\} = \text{gen}(\xi(F(\mathcal{J}_\tau)))$. □

We now want to apply this notion to several specific cases dealing with change of rings. To begin with, consider a finitely generated projective generator P of mod-R and let $P^* = \text{Hom}_R(P,R) \in$ R-mod. Let S be the endomorphism ring of P. Then (R,P^*,P,S) is a Morita context and so determines a Morita equivalence between the categories R-mod and S-mod. In particular, we have the inverse category isomorphisms

$$G = P \otimes_R -: \text{R-mod} \to \text{S-mod}$$

and

$$H = P^* \otimes_S -: \text{S-mod} \to \text{R-mod}.$$

Both G and H preserve exactness and direct sums and so the corresponding functions

G^o: S-tors $\rightarrow$ R-tors

and

H^o: R-tors $\rightarrow$ S-tors

define a bijective correspondence between R-tors and S-tors.

(9.2) PROPOSITION: Let G, H be the functors defined above and let $\tau \in$ R-tors. Then the following conditions on a left R-module M are equivalent:

(1) M is τ-injective.

(2) $G(M)$ is $H^o(\tau)$-injective.

PROOF: (1) $\Rightarrow$ (2): Let $\sigma = H^o(\tau)$ and let I be a σ-dense left ideal of S. If $\alpha: I \rightarrow G(M)$ is an S-homomorphism, then $H(\alpha): H(I) \rightarrow HG(M) \cong M$ is an R-homomorphism and $H(I)$ is τ-dense in $H(S)$. Therefore, there exists an R-homomorphism $\beta: H(S) \rightarrow M$ that extends $H(\alpha)$. Then $G(\beta): S \cong GH(S) \rightarrow G(M)$ is an S-homomorphism the restriction of which to I is α, proving that $G(M)$ is σ-injective.

(2) $\Rightarrow$ (1): This is proven similarly. □

(9.3) PROPOSITION: Let G, H be the functors defined above. Let $\tau \in$ R-tors and let $\sigma = H^o(\tau)$. If M is a left R-module then $G(E_\tau(M)) \cong E_\sigma(G(M))$.

PROOF: By Proposition 9.2, $G(E_\tau(M))$ is σ-injective. Since G induces an isomorphism between the submodule lattices of $E_\tau(M)$ and $G(E_\tau(M))$, it follows immediately that $G(E_\tau(M))$ is indeed a minimal σ-injective extension of $G(M)$ and so is isomorphic to $E_\sigma(G(M))$. □

A torsion theory $\tau \in R$-tors is said to be faithful if and only if $R \in \mathcal{F}_\tau$. These torsion theories will be discussed more fully in Section 10. Here we use only the fact that, for faithful τ, $Q_\tau(R) = E_\tau(R)$.

(9.4) PROPOSITION: *Let H be the functor defined above. Then $\tau \in R$-tors is faithful if and only if $H^\circ(\tau)$ is faithful.*

PROOF: Let $\sigma = H^\circ(\tau)$. If τ is faithful then $R \in \mathcal{F}_\tau$ and so, by the closure of $\mathcal{F}_\tau$ under taking direct products and submodules, any finitely generated projective left R-module belongs to $\mathcal{F}_\tau$. In particular, $P^* = H({}_S S) \in \mathcal{F}_\tau$ whence $S \in \mathcal{F}_\sigma$. Therefore, σ is faithful. The other direction is proven similarly. □

(9.5) PROPOSITION: *Let H, G be the functors defined above. Let $\tau \in R$-tors be faithful and let $\sigma = H^\circ(\tau) \in S$-mod. Then*

(1) *R_τ-mod and R_σ-mod are Morita equivalent;*

(2) *$P \otimes_R R_\tau$ is a finitely generated projective*

<u>generator of</u> mod-R_τ;

(3) $R_\sigma \cong \mathrm{Hom}_{R_\tau}(P \otimes_R R_\tau,\ P \otimes_R R_\tau)$.

PROOF: (1) By Proposition 2.3 and the fact that $G(P^*) \cong {}_S S$, we have $\mathrm{Hom}_R(E_\tau(P^*), E_\tau(P^*)) \cong \mathrm{Hom}_S(G(E_\tau(P^*)), G(E_\tau(P^*))) \cong \mathrm{Hom}_S(E_\sigma(G(P^*)), E_\sigma(G(P^*))) \cong \mathrm{Hom}_S(E_\sigma(S), E_\sigma(S))$. Since τ is faithful, so also is σ by Proposition 9.4 and thus $E_\sigma(S) = Q_\sigma(S)$ and we have $\mathrm{Hom}_S(E_\sigma(S), E_\sigma(S)) = S_\sigma$. Thus $\mathrm{Hom}_R(E_\tau(P^*), E_\tau(P^*)) \cong S_\sigma$.

Since P^* is a finitely generated projective left R-module, it is isomorphic to a direct summand of a finite direct sum of copies of R. Since it is a generator of R-mod, R is isomorphic to a direct summand of a finite direct sum of copies of P^*. Since the class of τ-injective left R-modules is closed under taking finite direct sums, it then follows that $E_\tau(R)$ and $E_\tau(P^*)$ are each isomorphic to a direct summand of a finite direct sum of copies of the other. By [72, Theorem 1.5] it then follows that $\mathrm{Hom}_R(E_\tau(R), E_\tau(R))$ is Morita equivalent to $\mathrm{Hom}_R(E_\tau(P^*), E_\tau(P^*))$. Since τ is faithful, $E_\tau(R) = Q_\tau(R)$ and so $R_\tau = \mathrm{Hom}_R(E_\tau(R), E_\tau(R))$, proving (1).

(2) Again by [72], $\mathrm{Hom}_R(E_\tau(P^*), E_\tau(R))$ is a finitely generated projective generator both of S_σ-mod and of mod-R_τ. By Proposition 9.2, $G(E_\tau(R)) = P \otimes_R E_\tau(R)$ is σ-injective and so from the exact sequence $0 \to S \to E_\sigma(S) \to E_\sigma(S)/S \to 0$

we derive the exact sequence $0 \to \mathrm{Hom}_S(E_\sigma(S)/S, P \otimes_R E_\tau(R)) \to \mathrm{Hom}_S(E_\sigma(S), P \otimes_R E_\tau(R)) \to \mathrm{Hom}_S(S, P \otimes_R E_\tau(R)) \to 0$. But $E_\sigma(S)/S$ is σ-torsion and $P \otimes_R E_\tau(R)$ is σ-torsion-free since $E_\tau(R)$ is τ-torsion-free so we have in fact $\mathrm{Hom}_R(E_\tau(P^*), E_\tau(R)) \cong \mathrm{Hom}_S(E_\sigma(S), P \otimes_R E_\tau(R)) \cong \mathrm{Hom}_S(S, P \otimes_R E_\tau(R)) \cong P \otimes_R E_\tau(R) \cong P \otimes_R R_\tau$, which proves that $P \otimes_R R_\tau$ is a finitely generated projective generator of mod-R_τ.

(3) This proof follows directly from (1) and (2). □

As another application of Proposition 9.1, we now consider a ring homomorphism $\gamma: R \to S$. Then γ induces functors

γ_*: S-mod → R-mod (restriction of scalars)

and

γ^*: R-mod → S-mod (extension of scalars),

respectively defined on objects by $\gamma_*({}_SN) = {}_RN$ and $\gamma^*({}_RM) = S \otimes_R M$. (The corresponding definitions for morphisms are obvious.) Then γ^* is the left adjoint of γ_*.

The functor γ_* is clearly exact and commutes with direct sums. It therefore defines a function $\gamma_\# = (\gamma_*)^\circ$: R-tors → S-tors, which assigns to each $\tau \in$ R-tors the torsion theory $\sigma \in$ S-tors characterized by $\mathcal{J}_\sigma = \{N \in \text{S-mod} \mid {}_RN \in \mathcal{J}_\tau\}$.

If S is flat as a right R-module, then γ^* is exact and commutes with direct sums. Thus it defines a function $\gamma^{\#} = (\gamma^*)^{\circ}$: S-tors → R-tors which assigns to each $\sigma \in$ S-tors the torsion theory $\tau \in$ R-tors characterized by $\mathcal{J}_{\tau} = \{M \in \text{R-mod} \mid S \otimes_R M \in \mathcal{J}_{\sigma}\}$.

(9.6) PROPOSITION: <u>Let</u> $\tau \in$ R-tors <u>and let</u> $\gamma: R \to S$ <u>be a ring homomorphism having the property that</u> $\gamma(R)$ <u>is</u> τ-<u>dense submodule of</u> $_RS$. <u>Let</u> $\tau' \in \text{gen}(\tau)$ <u>and let</u> $\sigma' = \gamma_{\#}(\tau')$. <u>Then</u> $\mathcal{L}_{\sigma'} = \{_SH \subseteq S \mid S\gamma(I) \subseteq H$ <u>for some</u> $I \in \mathcal{L}_{\tau'}\}$.

PROOF: By definition, $\mathcal{L}_{\sigma'} = \{_SH \subseteq S \mid {_R(S/H)} \in \mathcal{J}_{\tau'}\}$ Let $H \in \mathcal{L}_{\sigma'}$ and let $I = \gamma^{-1}(H)$. Then γ induces an R-monomorphism $R/I \to S/H$ and so $I \in \mathcal{L}_{\tau'}$. Furthermore, $H \supseteq S\gamma(I)$. Conversely, suppose that $I \in \mathcal{L}_{\tau'}$ and let $I' = \gamma^{-1}(S\gamma(I)) \supseteq I$. Then $I' \in \mathcal{L}_{\tau'}$ and we have the exact sequence of left R-modules $0 \to R/I' \overset{\overline{\gamma}}{\to} S/\gamma(I') \to S/\gamma(R) \to 0$, where $\overline{\gamma}$ is the R-homomorphism canonically induced by γ. By hypothesis, ${_RS/\gamma(R)} \in \mathcal{J}_{\tau} \subseteq \mathcal{J}_{\tau'}$ and so $\gamma(I')$ is τ'-dense in S. Therefore, $S/S\gamma(I') \in \mathcal{J}_{\tau'}$ whence $S\gamma(I) \in \mathcal{L}_{\sigma'}$. □

(9.7) PROPOSITION: <u>If</u> $\tau \in$ R-tors <u>then the function</u> $\hat{\tau}_{\#}: \text{gen}(\tau) \to R_{\tau}\text{-tors}$ <u>is monic</u>.

PROOF: Assume that $\tau' \neq \tau'' \in \text{gen}(\tau)$ and that

$\hat{\tau}_{\#}(\tau') = \hat{\tau}_{\#}(\tau'')$. Without loss of generality we can assume that there exists a nonzero $M \in \mathcal{J}_{\tau'} \setminus \mathcal{J}_{\tau''}$. Replacing M by $M/T_{\tau''}(M)$ if necessary, we can in fact assume that $M \in \mathcal{J}_{\tau'} \cap \mathcal{F}_{\tau''}$. Since $\tau'' \geq \tau$, M is also τ-torsion-free and so $\hat{\tau}_M: M \to Q_\tau(M)$ is an R-monomorphism. Since M is large in $Q_\tau(M)$, we then have $Q_\tau(M) \in \mathcal{F}_{\tau''}$. On the other hand, $Q_\tau(M)/M \in \mathcal{J}_\tau \subseteq \mathcal{J}_{\tau'}$, so, by the exactness of the sequence $0 \to M \to Q_\tau(M) \to Q_\tau(M)/M \to 0$, we have $Q_\tau(M) \in \mathcal{J}_{\tau'}$. But $\tau_{\#}(\tau') = \tau_{\#}(\tau'')$, implying that $Q_\tau(M) \in \mathcal{J}_{\tau''}$, which is a contradiction. □

In particular, suppose that I is a two-sided ideal of R and that γ is the canonical ring surjection $R \to R/I = S$. If $\tau \in R\text{-tors}$ and $\sigma = \gamma_{\#}(\tau)$ then as an immediate consequence of the definition we have $\mathcal{L}_\sigma = \{H/I \subseteq S \mid H \in \mathcal{L}_\tau\}$.

(9.8) PROPOSITION: Let I be a two-sided ideal of a ring R and let $\gamma: R \to R/I = S$ be the canonical ring surjection. If $\tau \in R\text{-tors}$ and $\sigma = \gamma_{\#}(\tau)$ then for every left S-module N,

(1) $T_\tau({}_RN) = T_\sigma({}_SN)$;

(2) If ${}_RN$ is τ-injective then ${}_SN$ is σ-injective.

PROOF: (1) This proof follows from the fact that every R-submodule of ${}_RN$ is also an S-submodule of ${}_SN$.

(2) Let $H/I \in \mathcal{L}_\sigma$. Then $[R/I]/[H/I] \cong R/H \in \mathcal{T}_\tau$ so $H \in \mathcal{L}_\tau$. If $\alpha: H/I \to N$ is an S-homomorphism then α induces an R-homomorphism $\alpha': H \to N$ defined by $\alpha': h \mapsto (h + I)\alpha$. Since N is τ-injective as a left R-module, there exists an R-homomorphism $\beta': R \to N$ extending α'. In particular, $I\beta' = 0$ so β' induces an S-homomorphism $\beta: S \to N$ which extends α. Therefore, N is σ-injective as a left S-module. □

(9.9) PROPOSITION: Let I be a two-sided ideal of R and let $\gamma: R \to R/I = S$ be the canonical ring surjection. If $\tau \in$ R-tors and $\sigma = \gamma_\#(\tau)$ then the following conditions are equivalent for a left S-module N:

(1) $Q_\tau({}_RN) = Q_\sigma({}_SN)$ (as R-modules or as S-modules).

(2) $Q_\sigma({}_SN)$ is τ-injective.

(3) $Q_\tau({}_RN)$ is a left S-module.

PROOF: (1) ⇒ (2), (3): trivially.

(2) ⇒ (1): By Proposition 9.8, $T_\tau({}_RN) = T_\sigma({}_SN)$ and so, in particular, $N/T_\tau({}_RN)$ is a left S-module. By passing to this module we can assume without loss of generality that N is τ-torsion-free. Then $T_\tau(Q_\sigma(N)) = T_\sigma(Q_\sigma(N)) = 0$ so N can be regarded as a submodule of $Q_\sigma(N)$. Also, $Q_\sigma(N)/N$ is σ-torsion and hence, as a left R-module, N is τ-dense in $Q_\sigma(N)$. By the τ-injectivity of

$Q_\tau({}_R N)$ there then exists a unique R-homomorphism α making the diagram

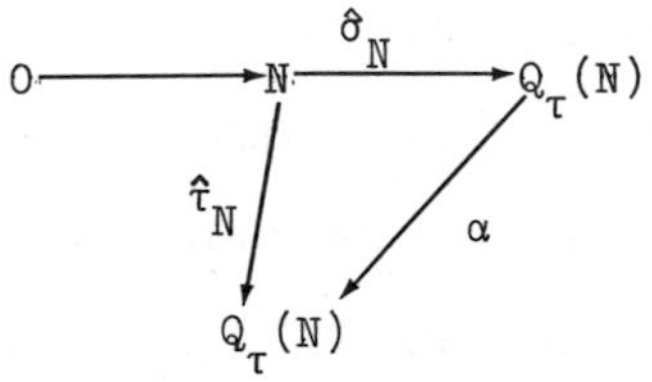

commute. Moreover, by (2) there exists a unique R-homomorphism β making the diagram

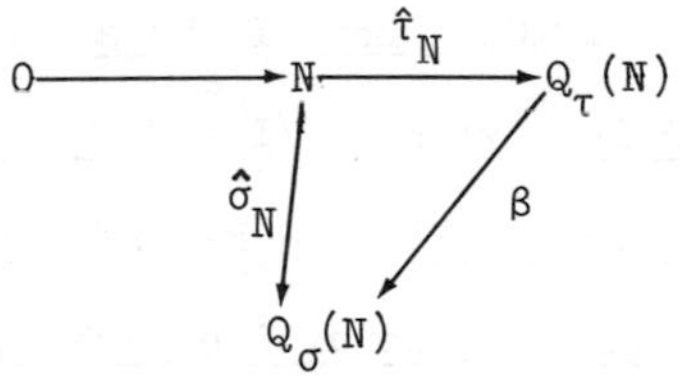

commute. Then by uniqueness $\alpha\beta$ and $\beta\alpha$ must be the respective identity maps and so α is an R-isomorphism. It is also clearly an S-isomorphism.

(3) ⇒ (1): As in the preceding part of the proof we can assume that $N \in \mathcal{F}_\tau$. Then $Q_\tau(N) \in \mathcal{F}_\sigma$ and is moreover σ-injective by Proposition 9.8. Also, N is σ-dense in $Q_\tau(N)$. The existence of an S-isomorphism between $Q_\tau(N)$ and $Q_\sigma(N)$ then follows as in the preceding paragraph, and this is clearly an R-isomorphism. □

Suppose in particular that $R = R_1 \times R_2$ is a direct product of rings, and let $\lambda_i\colon R_i \to R$ and $\pi_i\colon R \to R_i$

$(i = 1,2)$ be the canonical inclusion and projection maps, respectively. Then based on the preceding discussion we note the following:

(1) If $\sigma \in R_i$-tors then $(\lambda_i)_\#(\sigma) = \{N \in R\text{-mod} \mid {}_{R_i}N \in \mathcal{J}_\sigma\}$. Since each left R_i-module is also a left R-module, $(\lambda_i)_\#(\sigma) \neq \xi$ for any nontrivial $\sigma \in R_i$-tors.

(2) If $\sigma \in R_i$-tors then $(\pi_i)_\#(\lambda_i)(\sigma) = \sigma$.

Moreover, every left ideal I of R is of the form $I_1 \times I_2$ where I_i is a left ideal of R_i. Therefore, if $\sigma_i \in R_i$-tors and $\tau_i = (\lambda_i)_\#(\sigma_i)$ $(i = 1, 2)$ then $\mathcal{L}_{\tau_i} = \{I_1 \times I_2 \mid I_i \in \mathcal{L}_{\sigma_i}\}$. In particular, $\mathcal{L}_{\tau_1 \wedge \tau_2} = \{I_1 \times I_2 \mid I_1 \in \mathcal{L}_{\sigma_1}$ and $I_2 \in \mathcal{L}_{\sigma_2}\}$. We denote the torsion theory $\tau_1 \wedge \tau_2$ by $\sigma_1 \times \sigma_2$ and call it the <u>direct product</u> of σ_1 and σ_2.

<u>References for Section 9</u>

Cunningham [33]; Golan [56]; Raynaud [125]; Rubin [129]; Stenström [147]; Turnidge [160].

CHAPTER III

SPECIAL TORSION THEORIES

10. FAITHFUL TORSION THEORIES

Let $\tau \in$ R-tors. Recall that in Section 9 we called τ faithful if and only if $R \in \mathcal{F}_\tau$. From the definitions it is immediate that τ is faithful if and only if $\hat{\tau}: R \to R_\tau$ is a monomorphism. We then usually consider R as a subring of R_τ. Similarly, if $E \in \tau$ and B is the bicommutator of E, then by Proposition 6.11 τ is faithful if and only if $\delta_E: R \to B$ is a monomorphism.

(10.1) PROPOSITION: The family of all faithful $\tau \in$ R-tors is closed under taking joins.

PROOF: Let $U \subseteq$ R-tors and assume that each $\tau \in U$ is faithful. Then $R \in \mathcal{F}_\tau$ for all $\tau \in U$ and so $R \in \cap \mathcal{F}_\tau = \mathcal{F}_{\vee U}$. Thus $\vee U$ is faithful. □

In particular, by Proposition 10.1, R-prop has a unique maximal faithful torsion theory, which clearly must be $\chi(R)$. This theory, called the Lambek torsion theory, is important in the history of torsion theories and is discussed in

Chapter VII, Example 9.

It is natural for us to ask when every $\tau \in R$-prop is faithful, i.e. when $\chi(R)$ is the unique maximal element of R-prop. This is clearly true if R is a simple ring. In general, we first note the following result.

(10.2) PROPOSITION: The following conditions are equivalent for a ring R:

(1) Every $\tau \in R$-prop is faithful.

(2) Every $\sigma \in R_n$-prop is faithful for all $n \geq 1$.

(3) There exists an $n \geq 1$ such that for every $\sigma \in R_n$-prop is faithful.

PROOF: This is an immediate consequence of Proposition 9.4. □

(10.3) PROPOSITION: The following conditions are equivalent for a ring R:

(1) Every $\tau \in R$-prop is faithful.

(2) Every nonzero injective left R-module is faithful.

(3) If $0 \neq I$ is a proper left ideal of R then there exists an $a \in I$ and a $b \in R \setminus I$ with $(0:a) \subseteq (I:b)$.

(4) If $0 \neq I$ is a proper two-sided ideal of R then there exists an $a \in I$ and a $b \in R \setminus I$

<u>with</u> $(0:a) \subseteq (I:b)$.

PROOF: (1) ⇒ (2): Let E be a nonzero injective left R-module and let τ be the equivalence class of E. Since E is injective, any R-homomorphism $\alpha\colon (0:E) \to E$ can be extended to an R-homomorphism $\beta\colon R \to E$. In particular, if $x = 1\beta \in E$, then for any $r \in (0:E)$, we have $r\alpha = rx = 0$. Thus $\mathrm{Hom}_R((0:E),E) = 0$ and so $(0:E) \subseteq T_\tau(R) = 0$, proving that E is faithful.

(2) ⇒ (1): Let $\tau \in$ R-prop and let $0 \neq M$ be a τ-torsion-free left R-module. Then $\mathrm{Hom}_R(T_\tau(R),E(M)) = 0$ and so $T_\tau(R)E(M) = 0$. Since $E(M)$ is faithful, this means that $T_\tau(R) = 0$, proving (1).

(1) ⇒ (3): This follows from Proposition 2.4.

(3) ⇒ (4): This proof is trivial.

(4) ⇒ (1): This follows from Proposition 2.4, considering the fact that for any $\tau \in$ R-prop, the $T_\tau(R)$ is a two-sided ideal of R. □

(10.4) PROPOSITION: <u>If every</u> $\tau \in$ R-prop <u>is faithful then</u> R <u>has no nontrivial central idempotents</u>.

PROOF: Let e be a central idempotent of R and let $I = Re$. Then I is an idempotent two-sided ideal of R and so defines a $\tau \in$ R-tors by $\mathcal{L}_\tau = \{{}_RH \subseteq R \mid I \subseteq H\}$. If $(0:e) = 0$ then $e = 1$. Suppose, therefore, that

$(0:e) \neq 0$. Then there exists an $0 \neq a \in (0:e)$ and $I \subseteq (0:a)$. Thus $(0:a) \in \mathcal{L}_\tau$ which shows that τ is not faithful. By hypothesis, this means that $\tau = \chi$. Hence $I = 0$ and so $e = 0$. □

(10.5) PROPOSITION: <u>If every</u> $\tau \in$ R-prop <u>is faithful then</u> $R = \cap\{Q_\tau(R) \mid \tau \in \text{R-prop}\}$.

PROOF: Since each $\tau \in$ R-prop is faithful, $Q_\tau(R) = E_\tau(R) \subseteq E(R)$ so this intersection makes sense. If $x \in \cap\{Q_\tau(R) \mid \tau \in \text{R-prop}\}$ then $x + R \in T_\tau(E(R)/R)$ for every $\tau \in$ R-prop and in particular for $\tau = \chi([Rx + R]/R)$. This implies that $x + R = 0$ and so $x \in R$. □

Finally, we give a sufficient condition for a faithful torsion theory to be maximal.

(10.6) PROPOSITION: <u>A sufficient condition for a faithful</u> $\tau \in$ R-tors <u>to be equal</u> $\chi(R)$ <u>is that</u> R_τ <u>contain a completely-reducible subring</u> S <u>containing</u> R.

PROOF: Assume that such an S exists and let $I \in \mathcal{L}_{\chi(R)}$. By Proposition 1.7, I is a large left ideal of R. This implies that SI is a large left ideal of S and so $SI = S$, since a completely-reducible ring has no proper large left ideals. In particular, $1 = \Sigma s_i a_i$ for some $s_i \in S$ and $a_i \in I$. Let $H = \cap(R:s_i)$. Then $H \in \mathcal{L}_\tau$

since $(R:x) \in \mathcal{L}_\tau$ for any $x \in R_\tau$. But $H = H \cdot 1 = \Sigma H s_i a_i \subseteq I$ and so this implies that $I \in \mathcal{L}_\tau$. Therefore $\mathcal{L}_{\chi(R)} \subseteq \mathcal{L}_\tau$ and so $\chi(R) \leq \tau$ by Proposition 7.1. Since τ is faithful the reverse inequality holds and so $\tau = \chi(R)$. □

References for Section 10

Cunningham [33]; Goldman [63]; Teply [154].

11. STABLE TORSION THEORIES

Let $\tau \in R\text{-tors}$. We say that τ is <u>stable</u> if and only if $\mathcal{T}_\tau$ is closed under taking injective hulls. For example, the standard torsion theory over an integral domain and the Goldie torsion theory τ_G (see Chapter VII, Example 7) are both stable. Moreover, if $\tau \in \text{gen}(\tau_G)$ and M is a τ-torsion left R-module, then by Proposition 7.4 M is τ-dense in $E(M)$. Therefore, by the exactness of the sequence $0 \to M \to E(M) \to E(M)/M \to 0$ we have $E(M) \in \mathcal{T}_\tau$, proving that τ is stable.

(11.1) PROPOSITION: *A sufficient condition for $\tau \in R\text{-tors}$ to be stable is that every nonzero τ-torsion-free left R-module have a nonzero projective submodule.*

PROOF: By the preceding remarks it suffices to

show that if τ satisfies this condition then $\tau \geq \tau_G$. To do this we need to show that every large left ideal of R is contained in $\mathcal{L}_\tau$. Assume, therefore, that I is a large left ideal of R and that $H \neq R$ is the τ-purification of I in R. Then R/H is a nonzero τ-torsion-free left R-module and so R/H has a nonzero projective submodule K/H. By projectivity, we then have $K = H \oplus H'$ where $H' \cong K/H$. But this contradicts the fact that I, and hence H, is large in R. Therefore, we must have H = R, i.e. $I \in \mathcal{L}_\tau$. □

(11.2) PROPOSITION: <u>The following conditions are equivalent for</u> $\tau \in$ R-tors:

(1) τ <u>is stable</u>;

(2) $T_\tau(M)$ <u>is a direct summand of every injective left</u> R-<u>module</u> M;

(3) $T_\tau(M)$ <u>is a direct summand of every</u> τ-<u>injective left</u> R-<u>module</u> M.

PROOF: (1) ⇒ (2): If M is an injective left R-module then $E(T_\tau(M)) \subseteq M$. By (1), $E(T_\tau(M))$ is τ-torsion and so we must have $E(T_\tau(M)) = T_\tau(M)$. By injectivity, $T_\tau(M)$ is therefore a direct summand of M.

(2) ⇒ (3): Let M be a τ-injective left R-module. Then M is τ-pure in E(M) by Proposition 4.1, and so $T_\tau(E(M)) \subseteq M$. By (2), $T_\tau(E(M))$ is a direct summand of

$E(M)$ and so is injective. Therefore, $T_\tau(E(M))$ is a direct summand of M. Since $T_\tau(M) = T_\tau(E(M)) \cap M = T_\tau(E(M))$ by the left exactness of $T_\tau(\_)$, we thus have (3).

(3) ⇒ (1): If M is a τ-torsion left R-module, then by (3), $T_\tau(E(M))$ is a direct summand of $E(M)$. This direct summand must contain M, which is large in $E(M)$, and therefore must be all of $E(M)$. Thus $T_\tau(E(M)) = E(M)$, proving that $E(M)$ is τ-torsion. □

(11.3) PROPOSITION: *If* $\tau \in$ R-tors *is stable, then any indecomposable injective left* R-*module is either* τ-*torsion or* τ-*torsion-free. The converse holds if* R *is left noetherian*.

PROOF: Let $\tau \in$ R-tors be stable and let M be an indecomposable injective left R-module. By Proposition 11.2, $T_\tau(M)$ is a direct summand of M and so, by indecomposability, it must be either 0 or all of M, proving that M is either τ-torsion-free or τ-torsion.

Now assume that R is left noetherian and that every indecomposable injective left R-module is either τ-torsion or τ-torsion-free. Let $M \in \mathcal{T}_\tau$. Since R is left noetherian, $E(M) = \oplus E_i$ where the E_i are indecomposable injective left R-modules. Since M is large in $E(M)$, then $M \cap E_i \neq 0$ for each i and so $E_i \notin \mathcal{F}_\tau$ for each i.

Therefore, by hypothesis, $E_i \in \mathcal{J}_\tau$ for each i and so E(M) is τ-torsion. □

In Section 3 we remarked that, for a submodule N of a left R-module M, the topology $X_\tau(N)$ need not coincide with the restriction of the topology $X_\tau(M)$ to N. The next result shows that this does happen precisely when τ is stable.

(11.4) PROPOSITION: <u>The following conditions are equivalent for</u> τ ∈ R-tors:

(1) τ <u>is stable</u>.

(2) <u>If</u> N <u>is a submodule of a left</u> R-<u>module</u> M <u>then the restriction of</u> $X_\tau(M)$ <u>to</u> N <u>coincides with</u> $X_\tau(N)$.

PROOF: (1) ⇒ (2): Let N be a submodule of a left R-module M and let M' be a τ-dense submodule of N. We must find an open submodule M" of M such that $M'' \cap N = M'$. By Zorn's Lemma we can pick a submodule M" of M that is maximal with respect to the property $M'' \cap N = M'$. Then $N/M' = N/[M'' \cap N] \cong [N + M'']/M''$ which is large in M/M''. Since $N/M' \in \mathcal{J}_\tau$, (1) implies that $M/M'' \in \mathcal{J}_\tau$ and so M" is open in M. This proves (2).

(2) ⇒ (1): Note that M is τ-torsion if and only if $X_\tau(M)$ is the discrete topology. If $M \in \mathcal{J}_\tau$, then by (2)

the restriction of $X_\tau(E(M))$ to M coincides with the discrete topology, and so there exists an open submodule M' of $E(M)$ satisfying $M' \cap M = 0$. But by the largeness of M in $E(M)$ we must then have $M' = 0$. Therefore, $X_\tau(E(M))$ is also the discrete topology and so $E(M)$ is τ-torsion.

(11.5) COROLLARY: Let $\tau \in R\text{-tors}$ be stable and let M be a left R-module. If $M'' = \cap\{{}_RM' \subseteq M \mid M'$ is $X_\tau(M)$-open in $M\}$ then M'' has no proper τ-dense submodules.

PROOF: The proof follows directly from Proposition 11.4. □

(11.6) PROPOSITION: The family of all stable torsion theories in R-tors is closed under taking meets.

PROOF: Assume that $U \subseteq R\text{-tors}$ and that each $\tau \in U$ is stable. Let $M \in \mathcal{T}_{\wedge U}$. Then $M \in \mathcal{T}_\tau$ for every $\tau \in U$ and so, by stability, $E(M) \in \mathcal{T}_\tau$ for every $\tau \in U$. Therefore, $E(M) \in \cap \mathcal{T}_\tau = \mathcal{T}_{\wedge U}$. □

(11.7) PROPOSITION: The following conditions are equivalent for a stable $\tau \in R\text{-tors}$:

(1) $R \cong T_\tau(R) \times S$ where S is a completely reducible ring.

(2) Every τ-torsion-free left R-module is injective.

PROOF: (1) ⇒ (2): Assume that $R \cong T_\tau(R) \times S$ where S is a completely reducible ring. Then for any τ-torsion-free left R-module M, we have $T_\tau(R)M \subseteq T_\tau(M) = 0$. Let I be a left ideal of R. Then I can be written as $I_1 \oplus I_2$ where $I_1 \subseteq T_\tau(R)$ and $I_2 \subseteq S$. Also, $1 = a_1 + a_2$ where $a_1 \in T_\tau(R)$ and $a_2 \in S$.

Let M be a τ-torsion-free left R-module and let $\alpha\colon I \to M$ be an R-homomorphism. Then for every $b \in I_1$, $b\alpha = a_1(b\alpha) \in T_\tau(R)M = 0$. Let $\alpha'\colon I_2 \to M$ be the restriction of α to I_2. Then α' is an S-homomorphism and, by the complete reducibility of S, can be extended to an S-homomorphism $\beta'\colon S \to M$. Now define the R-homomorphism $\beta\colon R \to M$ by $\beta\colon a + s \mapsto s\beta'$. Then β extends α and so M is injective.

(2) ⇒ (1): If N is a submodule of a τ-torsion-free left R-module M, then N is τ-torsion-free and so, by (2), N is injective and hence a direct summand of M. Therefore, M is completely reducible. In particular, $\mathcal{F}_\tau$ is closed under taking homomorphic images.

Let M be a left R-module with $T_\tau(M) \neq 0$. If $T_\tau(M)$ is large in M then M is τ-torsion by stability. Otherwise, there exists a nonzero submodule N' of M that is maximal with respect to $T_\tau(M) \cap N' = 0$. Then N' is τ-torsion-free and so is injective, whence $M = N'' \oplus N'$ for some submodule N" of M. Since $M/N'' \cong N' \in \mathcal{F}_\tau$ we must have $T_\tau(M) \subseteq N''$. The maximality of N' implies that in fact $T_\tau(M)$ is large in N" and so, by stability, $T_\tau(M) = N''$. In particular, $R = T_\tau(R) \oplus N'$ where N' is a completely reducible left ideal of R (since N' is τ-torsion-free). If $r \in R$, then $T_\tau(R)r$ is τ-torsion and so $T_\tau(R)r \subseteq T_\tau(R)$, and N'r is τ-torsion-free since it is a homomorphic image of N' and so $N'r \subseteq N'$. Thus $R \cong T_\tau(R) \times N'$. □

References for Section 11

Armendariz [6]; Bernhard [14]; Gabriel [52]; Stenström [147]; Teply [154, 158].

12. SEMISIMPLE TORSION THEORIES

For a ring R we select a complete set of representatives of the isomorphism classes of simple left R-modules and denote it by R-simp. If τ ∈ R-tors we say that τ is semisimple if and only if $\tau = \xi(\mathcal{A})$ for some subset $\mathcal{A}$ of R-simp. If $\mathcal{A} = \{M\}$ we say that τ is simple.

If $\tau = \xi(\mathcal{A})$ is semisimple then it is easily seen that $M \in \mathcal{T}_\tau$ if and only if every homomorphic image of M has a nonzero submodule isomorphic to a member of $\mathcal{A}$. By Propositions 2.6 and 2.7, this holds if and only if $M = F_{k(M)}(M)$ where $\langle F_i \rangle$ is the $\mathcal{A}$-Loewy sequence on R-mod.

(12.1) PROPOSITION: Every nontrivial $\tau \in$ R-tors has a simple specialization. In particular, the simple torsion theories are the atoms of the lattice R-tors.

PROOF: Let $\tau \in$ R-tors be nontrivial. Then $\mathcal{J}_\tau$ contains a nonzero module and hence a nonzero cyclic module. This cyclic module in turn has a nonzero simple homomorphic image M, which also belongs to $\mathcal{J}_\tau$. Thus $\xi \neq \xi(M) \leq \tau$. If τ is an atom of R-tors, then we must have $\xi(M) = \tau$. □

In particular, if M and M' are nonisomorphic simple left R-modules, then $\xi(M)$ and $\xi(M')$ are distinct torsion theories. By Proposition 12.1, it follows that in fact $\xi(M)$ and $\xi(M')$ are disjoint.

(12.2) PROPOSITION: If M and M' are nonisomorphic simple left R-modules then $\xi(M) \leq \chi(M')$.

PROOF: Since $M' \notin \mathcal{J}_{\xi(M)}$ we must have $M' \in \mathcal{F}_{\xi(M)}$ and so, by Proposition 8.5, $\chi(M') \geq \xi(M)$. □

(12.3) PROPOSITION: If $\mathcal{A}$ is a subset of R-simp then $\wedge\{\chi(M) \mid M \in \mathcal{A}\} = \xi$ if and only if $\mathcal{A}$ = R-simp.

PROOF: Let $\tau = \wedge\{\chi(M) \mid M \in \mathcal{A}\}$. If $\mathcal{A}$ = R-simp and $\tau \neq \xi$, then by Proposition 12.1 there exists an

$M \in$ R-simp with $\xi(M) \leq \tau$. In particular, $M \in \mathcal{J}_{\xi(M)} \subseteq \mathcal{J}_{\chi(M)}$, which is a contradiction. Conversely, if $\mathcal{A} \neq$ R-simp and $M \in$ R-simp $\setminus \mathcal{A}$, then by Proposition 12.2 $\xi(M) \leq \chi(M')$ for each $M' \in \mathcal{A}$ and so $\xi(M) \leq \tau$. Therefore $\tau \neq \xi$. □

It follows in particular that $\chi(M) = \xi$ if and only if M contains an isomorphic copy of every simple left R-module. Such modules are called <u>lower distinguished</u>.

(12.4) PROPOSITION: *A torsion theory* $\tau \in$ R-tors *is semisimple if and only if it is the join of simple torsion theories.*

PROOF: If $\tau = \xi(\mathcal{A})$ is semisimple then $\tau = \vee\{\xi(M) \mid M \in \mathcal{A}\}$ and each such $\xi(M)$ is simple. Conversely, suppose that $\tau = \vee\tau_i$ where the τ_i are simple torsion theories. Let $\mathcal{A} =$ R-simp $\cap\, \mathcal{J}_\tau$. Then $\xi(\mathcal{A}) \leq \tau$. On the other hand, let M be τ-torsion and let $N = T_{\xi(\mathcal{A})}(M)$. If $N \neq M$ then $M/N \in \mathcal{J}_\tau = \cap \mathcal{J}_{\tau_i}$, and so M/N has a simple submodule M' isomorphic to a member of $\mathcal{A}$. But M/N is $\xi(\mathcal{A})$-torsion-free, so $M' \in \mathcal{F}_{\xi(\mathcal{A})}$, which is a contradiction. Therefore, $N = M$ and so $\tau \leq \xi(\mathcal{A})$, proving equality. □

(12.5) PROPOSITION: *The following conditions are equivalent:*

(1) *If* $M \in$ R-simp *then* $E(M) \in \mathcal{J}_{\xi(M)}$.

(2) <u>Every left ideal of</u> R <u>can be written as the intersection of members of the</u> $\mathcal{L}_{\xi(M)}$, $M \in$ R-simp.

PROOF: (1) ⇒ (2): Let I be a left ideal of R. To show (2) it suffices to show that for every $r \in R \setminus I$ there exists an $M \in$ R-simp and an $H \in \mathcal{L}_{\xi(M)}$ such that $I \subseteq H$ and $r \notin H$. Indeed, let $\mathcal{A}$ be the set of all left ideals of R containing I but not r. Then $\mathcal{A}$ is nonempty since $I \in \mathcal{A}$. By Zorn's Lemma, $\mathcal{A}$ has a maximal element H'. Then $[Rr + H']/H'$ is a large simple submodule of R/H' and so is isomorphic to some $M \in$ R-simp. By (1), R/H' then belongs to $\mathcal{J}_{\xi(M)}$ and so $H' \in \mathcal{L}_{\xi(M)}$.

(2) ⇒ (1): We first claim that if I is a left ideal of R such that R/I contains a large simple submodule isomorphic to $M \in$ R-simp, then $I \in \mathcal{L}_{\xi(M)}$. Indeed, let $r \in R$ be chosen so that $[Rr + I]/I \cong M$. By (2), there then exists an $H \in \mathcal{L}_{\xi(M')}$ for some $M' \in$ R-simp with $I \subseteq H$ and $r \notin H$. But since $[Rr + I]/I$ is large in R/I, it then follows that $H = I$. Hence $M \cong M'$ and so $I \in \mathcal{L}_{\xi(M)}$.

Now let $M \in$ R-simp. If $0 \neq x \in E(M)$, then Rx contains M as a large simple submodule and so, by the above, $(0:x) \in \mathcal{L}_{\xi(M)}$, whence Rx is $\xi(M)$-torsion. Therefore, $E(M) \in \mathcal{J}_{\xi(M)}$, proving (1). □

A left R-module M is said to be <u>semiartinian</u> if and only if $M \in \mathcal{J}_{\xi(R\text{-simp})}$, that is to say, if and only if every nonzero homomorphic image of M has a nonzero socle. A ring R is said to be <u>left semiartinian</u> if and only if R is semiartinian as a left module over itself. This is equivalent to saying that every left R-module is semiartinian or, in other words, $\xi(R\text{-simp}) = \chi$.

(12.6) PROPOSITION: *The following conditions are equivalent for a ring* R:

(1) R *is left semiartinian*.

(2) *Every nontrivial* $\tau \in$ R-tors *is semisimple.*

(3) *The lattice* R-tors *is boolean*.

PROOF: (1) ⇒ (2): Suppose that R is left semiartinian. Let $\tau \in$ R-tors be nontrivial and let $\mathcal{A} = R\text{-simp} \cap \mathcal{J}_\tau$. Then $\xi(\mathcal{A}) \leq \tau$. On the other hand, if M is τ-torsion and $N = T_{\xi(\mathcal{A})}(M)$ then $0 \neq M/N \in \mathcal{J}_\tau$ implies that there exists a simple submodule M' of M/N that is τ-torsion. Therefore M' is isomorphic to a member of $\mathcal{A}$, and so $T_{\xi(\mathcal{A})}(M/N) \neq 0$, which is a contradiction. Hence $\tau \leq \xi(\mathcal{A})$ and so we have equality.

(2) ⇒ (3): Let $\tau \in$ R-prop be nontrivial. By (2), $\tau = \xi(\mathcal{A})$ for some subset $\mathcal{A}$ of R-tors. Let $\tau' = \xi(R\text{-simp} \setminus \mathcal{A})$. Then clearly $\tau \wedge \tau' = \xi$ and, by Proposition 12.4, $\tau \vee \tau' = \chi$. Thus R-tors is complemented.

By Proposition 8.11, it is distributive and so is boolean.

(3) ⇒ (1): Suppose that R-tors is boolean and that ξ(R-simp) is proper. Then there exists a $\tau \in$ R-prop such that $\tau \wedge \xi(\text{R-simp}) = \xi$ and $\tau \vee \xi(\text{R-simp}) = \chi$. By Proposition 12.1, τ has a simple specialization τ'. But then $\tau' \leq \xi(\text{R-simp}) \wedge \tau$, which is a contradiction. □

(12.7) COROLLARY: _If_ R _is a right perfect ring then every nontrivial_ $\tau \in$ R-tors _is semisimple_.

PROOF: Every right perfect ring is left semiartinian [9]. □

(12.8) PROPOSITION: _If_ R _is left semiartinian and left noetherian, then_ R _is left artinian_.

PROOF: Let $<F_i>$ be the Loewy sequence on R-mod, and for each ordinal i let $H_i = F_i(R)$. Then $H_1 \subseteq H_2 \subseteq \ldots$ is an ascending chain of left ideals of R and so, by left noetherianness, there exists a positive integer k with $H_k = H_{k+1} = \ldots$. Since R is a left semiartinian, we must have $H_k = R$. Thus R/H_{k-1} is a finite direct sum of simple left R-modules and so R has finite length, which suffices to show that R is left artinian. □

An ideal I of a ring R is said to be _right T-nilpotent_ if and only if, for every sequence $a_1, a_2, \ldots$

of elements of I, there exists a positive integer n for which $a_n a_{n-1} \cdot \ldots \cdot a_1 = 0$.

(12.9) PROPOSITION: The following conditions are equivalent for a ring R:

(1) R is left semiartinian.

(2) $\mathrm{soc}(M)$ is large in M for every $M \in R$-mod.

(3) (i) $J(R)$ is right T-nilpotent; and

(ii) $R/J(R)$ is left semiartinian.

(4) The functor $T_{\xi(R\text{-simp})}(\_)$ is exact.

PROOF: (1) ⇔ (2): The proof is trivial.

(1) ⇒ (3): Let $\langle F_i \rangle$ be the Loewy sequence on R-mod. Since R is left semiartinian, $J(R)$ is semiartinian as a left R-module and so there exists an ordinal k with $J(R) = F_k(J(R))$. For any $a \in J(R)$, define $h(a)$ to be the smallest ordinal i with $a \in F_i(R)$. Then $h(a)$ is clearly not a limit ordinal. Moreover, for any ordinal i, $J(R)F_i(R) \subseteq F_{i-1}(R)$, so for any $b \in J(R)$, $h(ba) < h(a)$. Now let $a_1, a_2, \ldots$ be a sequence of elements of $J(R)$ with $a_n a_{n-1} \cdot \ldots \cdot a_1 \neq 0$ for all positive integers n. Then we obtain an infinite descending sequence of ordinals $h(a_1) > h(a_2 a_1) > \ldots$, which cannot happen. Therefore, $J(R)$ is right T-nilpotent, proving 3(i). The proof of 3(ii) from (1) is immediate.

(3) ⇒ (1): Let M be a left R-module satisfying the condition that $J(R)N \neq 0$ for every nonzero submodule N of M. Then in particular there exists an $a_1 \in J(R)$ with $a_1M \neq 0$. Since $Ra_1M \neq 0$ there exists by the same reasoning an $a_2' \in J(R)$ with $a_2'Ra_1M \neq 0$, whence there exists an $a_2 \in J(R)$ with $a_2a_1M \neq 0$. Continuing in this manner we can obtain a sequence $a_1, a_2, \ldots$ of elements of $J(R)$ satisfying $a_na_{n-1} \cdot \ldots \cdot a_1M \neq 0$ for every positive integer n. This contradicts 3(i). We therefore conclude that every left R-module M has a nonzero submodule N satisfying $J(R)N = 0$. Then N is also a left $R/J(R)$-module, and so by 3(ii) has a nonzero simple submodule. This shows that M has a nonzero simple submodule and so proves (1).

(1) ⇒ (4): By (1), $\mathcal{T}_{\xi(R\text{-simp})} = \{0\}$ and so $\mathcal{T}_{\xi(R\text{-simp})}$ is trivially closed under taking homomorphic images. Then (4) follows from Proposition 5.6.

(4) ⇒ (1): Suppose that $R \notin \mathcal{J}_{\xi(R\text{-simp})}$, and let I be a maximal left ideal of R containing $T_{\xi(R\text{-simp})}(R)$. Then R/I is $\xi(R\text{-simp})$-torsion since it is simple, and is $\xi(R\text{-simp})$-torsion-free by Proposition 5.5 since it is a homomorphic image of $R/T_{\xi(R\text{-simp})}(R)$. This contradiction proves (1). □

(12.10) PROPOSITION: _If_ R _is left semiartinian then any_ $\tau \in$ R-prop _can be written as_

$\wedge\{\chi(M) \mid M \in \mathcal{A}\}$ <u>for some</u> $\mathcal{A} \subseteq$ R-simp.

PROOF: Let $\tau \in$ R-prop and let $\mathcal{A}$ = R-simp $\cap\, \mathcal{F}_\tau$. Then $\chi(M) \geq \tau$ for every $M \in \mathcal{A}$ and so $\wedge\chi(M) \geq \tau$. Now assume that $0 \neq N \in \mathcal{J}_{\wedge\chi(M)} \setminus \mathcal{J}_\tau$. Replacing N by $N/T_\tau(N)$ if necessary, we can in fact assume that N is τ-torsion-free. Since R is left semiartinian, N has a simple submodule N'. Since $N \in \mathcal{J}_{\wedge\chi(M)}$, N' is not isomorphic to a member of $\mathcal{A}$ and so N' is not τ-torsion-free. But by simplicity this implies that N' is τ-torsion, contradicting the fact that N is τ-torsion-free. Therefore $\tau = \wedge\chi(M)$. □

(12.11) PROPOSITION: <u>If</u> R <u>is left semiartinian and if</u> $\chi(M)$ <u>is stable for every</u> $M \in$ R-simp, <u>then every</u> $\tau \in$ R-prop <u>is stable</u>.

PROOF: This follows from Propositions 12.10 and 11.6. □

We will say that a ring R is <u>left local</u> if and only if R-tors has a unique atom. By Proposition 12.1, this is equivalent to saying that all simple left R-modules are isomorphic. If R is left local and I is a two-sided ideal of R, then I is contained in a maximal left ideal H of R. Moreover, $I = (0:R/I) \subseteq (0:R/H) = (0:M)$ where M is a representative of the isomorphism class of simple left ideals of R. Therefore, if R is left local it has a unique maximal two-sided ideal, namely $(0:M)$, and indeed $(0:M) = J(R)$.

(12.12) PROPOSITION: *If R is a left local ring then $J(R)$ is either 0 or large in R.*

PROOF: Let M be a representative of the isomorphism class of simple left R-modules. Then $J(R) = (0:M)$ is the unique maximal two-sided ideal of R. If I is a nonzero left ideal of R with $I \cap J(R) = 0$, then $IR = R$ by the maximality of $J(R)$, and so $J(R) \subseteq J(R)R = J(R)IR \subseteq [J(R) \cap I]R = 0$. If there is no such I then $J(R)$ is large in R. □

We now put together the various concepts introduced in this section to characterize those rings for which the lattice R-tors is as small as possible.

(12.13) PROPOSITION: *The following conditions are equivalent for a ring R:*

(1) R-tors $= \{\xi, \chi\}$.

(2) R *is left semiartinian and left local.*

(3) $R = S_n$ *where* S *is a left semiartinian ring having a unique maximal left ideal.*

(4) *Every* $\tau \in$ R-prop *is faithful and* $R \notin \mathcal{T}_{\xi}(\text{R-simp})$.

PROOF: (1) ⇒ (2): By (1), χ is an atom of R-tors and so $\chi = \xi(M)$ for some simple left R-module M. Therefore, $R \in \mathcal{J}_{\xi(M)}$ is left semiartinian. If M' is another simple left R-module, then $\xi(M') \neq \xi$. Thus $\xi(M') = \chi = \xi(M)$, and so by Proposition 12.2, $M \cong M'$. Thus R is left local.

(2) ⇒ (3): Since R is left semiartinian, then by Proposition 12.9, J(R) is right T-nilpotent and $\overline{R} = R/J(R)$ is left semiartinian. We claim that $\overline{R}$ is in fact completely reducible. Indeed, assume that $soc(\overline{R}) \neq \overline{R}$ and let I_0 be a maximal left ideal of $\overline{R}$ containing $soc(\overline{R})$. By Proposition 12.9, $soc(\overline{R})$ is large in $\overline{R}$ and so I_0 is large in $\overline{R}$.

On the other hand, let K be a simple left ideal of $\overline{R}$. Since $J(\overline{R}) = 0$, there exists a maximal left ideal I_1 of $\overline{R}$ with $K \cap I_1 = 0$ and so I_1 is not large in $\overline{R}$. Therefore, $K \cong R/I_1$ is not isomorphic to $\overline{R}/I_0$. But by (2) there exists only one isomorphism class of simple left R-modules and so we have a contradiction. Therefore, $\overline{R}$ is completely reducible.

By [9, Theorem P], R is thus right perfect and so, in particular, idempotents can be lifted modulo J(R). Hence $R \cong H^n$ where H is a left R-module having a unique maximal submodule H'. Thus the endomorphism ring S of H has a unique maximal left ideal $\{\beta \in S \mid H\beta \subseteq H'\}$. Then $R = S_n$, proving (3).

(3) ⇒ (1): If R is characterized as in (3), then it is clear that R-simp = {M} for some simple left R-module M and $\chi = \xi(\text{R-simp}) = \xi(M)$. Therefore, χ is an atom of R-tors and so by Proposition 12.1 we have R-tors = $\{\xi, \chi\}$.

(1), (2) ⇒ (4): The proof is trivial.

(4) ⇒ (2): Let I be a simple left ideal of R. By (4), $\xi(I)$ is faithful if $\xi(I) \neq \chi$, which would imply that $\mathrm{Hom}_R(I,R) \neq 0$, which is a contradiction. Therefore, $\xi(I) = \chi$ and so $\mathrm{Hom}_R(I,M) \neq 0$ for every nonzero left R-module M. This clearly implies (2). □

Recall that a ring R is _left primitive_ if and only if there exists a faithful simple left R-module. If R is left local and if M is a representative of the isomorphism class of simple left R-modules, then M is a faithful left $R/J(R)$-module (since $J(R) = (0:M)$), and so $R/J(R)$ is a left primitive ring.

(12.14) COROLLARY: _The following conditions are equivalent for a left primitive ring_ R:

(1) R _is left semiartinian and left local_.

(2) $R \cong S_n$ _where_ S _is a division ring_.

PROOF: (1) ⇒ (2): Since R is left local, all simple left R-modules are isomorphic and hence faithful.

Therefore $J(R) = 0$. The proof now follows the proof of (2) ⇒ (3) of Proposition 12.13.

(2) ⇒ (1): This follows from Proposition 12.13. □

In particular, if R is left primitive, left semiartinian, and left local, then R is a simple left artinian ring by the Wedderburn-Artin Theorem.

References for Section 12

Albu [2]; Alin [3]; Bican, Jambor, Kepka, and Němec [17]; Bronowitz and Teply [23]; Dickson [35, 36, 37, 38]; Gardner [53]; Nastasescu [110]; Nastasescu and Popescu [112]; Shores [139, 140, 141]; Teply [155].

13. SATURATED TORSION THEORIES

In the previous section we noted that the atoms of R-tors are precisely the simple torsion theories. In this section we would like to characterize the coatoms of R-tors (i.e., the maximal elements of R-prop). To do this we have to introduce the notion of a saturated torsion theory.

Let $\tau \in$ R-tors. Associated with τ we have the torsion theory $\bar{\tau} = \chi(R/T_\tau(R))$, called the saturation of τ. If $\bar{\tau} = \tau$ then we say that τ is saturated.

(13.1) PROPOSITION: The following conditions are equivalent for $\tau \in$ R-tors:

(1) τ is saturated.

(2) $\{\tau' \in R\text{-tors} \mid T_{\tau'}(R) = T_{\tau}(R)\} \subseteq \mathrm{spcl}(\tau)$.

PROOF: (1) ⇒ (2): If $T_{\tau'}(R) = T_{\tau}(R)$ then $R/T_{\tau}(R) \in \mathcal{F}_{\tau'}$, and so $\tau = \chi(R/T_{\tau}(R)) \geq \tau'$.

(2) ⇒ (1): To prove (1) it suffices to show that $T_{\bar{\tau}}(R) = T_{\tau}(R)$. Since $\tau \leq \bar{\tau}$, clearly $T_{\tau}(R) \subseteq T_{\bar{\tau}}(R)$. On the other hand, $R/T_{\tau}(R) \in \mathcal{F}_{\bar{\tau}}$ and so $T_{\bar{\tau}}(R) \subseteq T_{\tau}(R)$.

(13.2) PROPOSITION: <u>If</u> $\tau \in R\text{-tors}$ <u>is saturated then</u> $T_{\tau}(R) = (0:E(R/T_{\tau}(R)))$.

PROOF: Let $E = E(R/T_{\tau}(R))$. Then $\tau = \bar{\tau} = \chi(E)$. Since $T_{\tau}(R) \in \mathcal{J}_{\tau}$, we have $\mathrm{Hom}_R(T_{\tau}(R),E) = 0$ and in particular $T_{\tau}(R)E = 0$. Thus $T_{\tau}(R) \subseteq (0:E)$. Conversely, if $\alpha \in \mathrm{Hom}_R((0:E),E)$ then, by the injectivity of E, α can be extended to an R-homomorphism $\beta: R \to E$. If $x = 1\beta \in E$ then $\alpha: a \mapsto ax = 0$, so $\alpha = 0$. Thus $\mathrm{Hom}_R((0:E),E) = 0$, so $(0:E) \in \mathcal{J}_{\tau}$. This shows that $(0:E) \subseteq T_{\tau}(R)$ and so we have equality. □

(13.3) PROPOSITION: <u>The following conditions are equivalent for</u> $\tau \in R\text{-prop}$:

(1) τ <u>is a coatom of</u> R-tors.

(2) (i) τ <u>is saturated; and</u>

(ii) $T_{\tau'}(R) = T_{\tau}(R)$ <u>for all proper</u> $\tau' \in \mathrm{gen}(\tau)$.

PROOF: (1) ⇒ (2): If $\tau \in R\text{-prop}$ then

$\tau \leq \bar{\tau} \in$ R-prop, and so by (1), $\tau = \bar{\tau}$. Therefore, τ is saturated. If $\tau' \in \text{gen}(\tau)$ is proper then $\tau = \tau'$ by (1) and so surely $T_{\tau'}(R) = T_\tau(R)$.

(2) ⇒ (1): This follows directly from Proposition 13.1. □

(13.4) PROPOSITION: <u>If</u> R <u>satisfies the ascending the chain condition on two-sided ideals, then every</u> $\tau \in$ R-prop <u>has a generalization that is a coatom of</u> R-tors.

PROOF: If $\tau \in$ R-prop, let $\mathcal{A}$ be the set of all two-sided ideals of R that are τ-pure in R as left ideals. Then $\mathcal{A} \neq \emptyset$ since $T_\tau(R) \in \mathcal{A}$. By hypothesis, $\mathcal{A}$ therefore has a maximal element I. Since $R/I \in \mathcal{F}_\tau$, we have $\chi(R/I) \geq \tau$. If $\chi > \tau' \geq \chi(R/I)$ then $R/T_{\tau'}(R) \in \mathcal{F}_{\tau'} \subseteq \mathcal{F}_\tau$, and so by the maximality of I we have $T_{\tau'}(R) = I$. Thus $\chi(R/I)$ is saturated and indeed is a coatom of R-tors by Proposition 13.3. □

(13.5) PROPOSITION: <u>Let</u> $\tau \in$ R-prop <u>be saturated. A sufficient condition for</u> τ <u>to be a coatom of</u> R-tors <u>is that</u> $T_\tau(R)$ <u>be a prime ideal of</u> R. <u>If</u> $T_\tau(R)$ <u>is a semiprime ideal of</u> R <u>this condition is also necessary</u>.

PROOF: Let $\tau \in$ R-prop be saturated and let $T_\tau(R)$ be a prime ideal of R. Assume, furthermore, that

$\chi > \tau' \geq \tau$. If $I = T_{\tau'}(R)$ then $R/I \in \mathcal{F}_{\tau'} \subseteq \mathcal{F}_{\tau} = \mathcal{F}_{\bar{\tau}}$, and so there exists a nonzero R-homomorphism $\alpha: R/I \to E(R/T_{\tau}(R))$. In particular, $H/T_{\tau}(R) = \mathrm{im}(\alpha) \cap R/T_{\tau}(R)$ must be nonzero. Since $I = (0{:}R/I)$, then $I(\mathrm{im}(\alpha)) = 0$ and so $IH \subseteq T_{\tau}(R)$. Therefore, $I(HR) \subseteq T_{\tau}(R)$. Since $T_{\tau}(R)$ is prime, this implies that $I \subseteq T_{\tau}(R)$. But then $T_{\tau}(R)/I$ is both τ-torsion and τ-torsion-free and so $I = T_{\tau}(R)$. By Proposition 13.3, τ is therefore a coatom of R-tors.

Conversely, assume that $T_{\tau}(R)$ is a semiprime ideal of R and that τ is a coatom of R-tors. Let I be a two-sided ideal of R containing $T_{\tau}(R)$ which is τ-pure as a left ideal of R. Then $\chi(R/I) \geq \tau$ so, in particular, $T_{\tau}(R)$ is $\chi(R/I)$-torsion.

We claim that $I/T_{\tau}(R)$ is also $\chi(R/I)$-torsion. Indeed, let $\alpha: I/T_{\tau}(R) \to E(R/I)$ be a nonzero R-homomorphism. Since $E(R/I) \in \mathcal{F}_{\tau} = \mathcal{F}_{\bar{\tau}}$, we have $\mathrm{im}(\alpha) \in \mathcal{F}_{\bar{\tau}}$ and so there exists an R-homomorphism $\beta: E(R/I) \to E(R/T_{\tau}(R))$ with $\alpha\beta \neq 0$. Thus there exists a left ideal H of R, $I \supseteq H \supset T_{\tau}(R)$, for which $[H/T_{\tau}(R)]\alpha \subseteq R/I$ and $0 \neq [H/T_{\tau}(R)]\alpha\beta \subseteq R/T_{\tau}(R)$. Let $H'/T_{\tau}(R) = [H/T_{\tau}(R)]\alpha\beta$. Since $H \subseteq I$, then $H[H/T_{\tau}(R)]\alpha \subseteq H[R/I] = 0$ and so $HH' \subseteq T_{\tau}(R)$. Since $T_{\tau}(R)$ is a semiprime ideal of R, then $H'H \subseteq T_{\tau}(R)$ and so $H'[H/T_{\tau}(R)]\alpha\beta \subseteq [H'H/T_{\tau}(R)]\alpha\beta = 0$. In particular, $(H')^2 \subseteq T_{\tau}(R)$. Since $T_{\tau}(R)$ is semiprime, then $H' \subseteq T_\tau(R)$, which is a contradiction. Therefore, we must have $\text{Hom}_R(I/T_\tau(R), E(R/I)) = 0$ and so $I/T_\tau(R)$ is $\chi(R/I)$-torsion.

From the exact sequence $0 \to T_\tau(R) \to I \to I/T_\tau(R) \to 0$ and the preceding remarks, we deduce that I is $\chi(R/I)$-torsion and so $I \subseteq T_{\chi(R/I)}(R)$. But since τ is a coatom of R-tors, $T_{\chi(R/I)}(R) = T_\tau(R)$ by Proposition 13.3 and $T_\tau(R) \subseteq I$ by the choice of I. Therefore, $I = T_\tau(R) = T_{\chi(R/I)}(R)$. We therefore see that no two-sided ideal of R properly containing $T_\tau(R)$ is τ-pure as a left ideal of R.

Now let H and K be two-sided ideals of R with $HK \subseteq T_\tau(R)$ and $K \supset T_\tau(R)$. Then $\chi(K/T_\tau(R)) \geq \chi(R/T_\tau(R)) = \tau$ and so $K/T_\tau(R) \in \mathcal{F}_\tau$. In particular, $(0:K/T_\tau(R))$ is a two-sided ideal of R which is τ-pure as a left ideal of R and which contains $T_\tau(R)$. By the above preceding argument, $(0:K/T_\tau(R)) = T_\tau(R)$. But $H \subseteq (0:K/T_\tau(R))$ and so $H \subseteq T_\tau(R)$, proving that $T_\tau(R)$ is a prime ideal of R. □

(13.6) PROPOSITION: <u>Let</u> $\tau \in$ R-tors <u>be saturated. Then the following conditions are equivalent for a</u> τ<u>-torsion-free left</u> R-<u>module</u> M:

(1) M <u>is injective as a left</u> R-<u>module</u>.

(2) M <u>is injective as a left</u> R_τ-<u>module</u>.

(3) M <u>is injective as a left</u> $R/T_\tau(R)$-<u>module</u>.

PROOF: Let $E = E(R/T_\tau(R))$. Since τ is saturated, $E \in \tau$. By Proposition 13.2, E is a left $R/T_\tau(R)$-module and so E is the $R/T_\tau(R)$-injective hull of $R/T_\tau(M)$. Since $E \in \mathcal{E}_\tau$, E is also a left R_τ-module and is injective by Proposition 6.7. Then each of the given conditions is equivalent to the assertion that M is a direct summand of a direct product of copies of E. □

(13.7) PROPOSITION: The following conditions are equivalent for $\tau \in$ R-tors:

(1) τ is a coatom of R-tors.

(2) (i) τ is saturated; and

(ii) every nonzero injective left R_τ-module that is τ-torsion-free as a left R-module is faithful over R_τ.

(3) (i) τ is saturated; and

(ii) every nonzero injective left $R/T_\tau(R)$-module that is τ-torsion-free as a left R-module is faithful over $R/T_\tau(R)$.

PROOF: (1) ⇒ (2): That (1) implies (2i) follows from Proposition 13.3. Let M be a nonzero injective left R_τ-module that is τ-torsion-free as a left R-module. By Proposition 13.6, M is injective as a left R-module. Since M is τ-torsion-free, $\chi > \chi(M) \geq \tau$ and so, by (1), $\chi(M) = \tau$. In particular, $R_\tau \in \mathcal{F}_{\chi(M)}$ and so R_τ can be embedded in a direct product of copies of M. This shows that M is faithful as a left R_τ-module.

(2) ⇒ (3): This follows from Proposition 13.6 and the fact that $R/T_\tau(R) \subseteq R_\tau$.

(3) ⇒ (1): If $\tau' \in$ R-prop with $\tau' \geq \tau$, then $E = E(R/T_{\tau'}(R))$ is injective and τ-torsion-free and so it is an injective $R/T_\tau(R)$-module. By (3), E is then faithful. Thus $T_{\tau'}(R) = (0:E(R/T_{\tau'}(R))) = T_\tau(R)$ and so (1) follows by Proposition 13.3. □

References for Section 13

Beachy [11]; Popescu and Spircu [123].

14. TORSION THEORIES SATISFYING CHAIN CONDITIONS

Let $\tau \in$ R-tors. We say that τ is noetherian if and only if, for every ascending chain $I_1 \subseteq I_2 \subseteq \cdots$ of left ideals of R with $\cup I_j \in \mathcal{L}_\tau$, there exists a positive integer k for which $I_k \in \mathcal{L}_\tau$.

(14.1) PROPOSITION: The following conditions are equivalent for $\tau \in$ R-tors:

(1) τ is noetherian.

(2) $\mathcal{E}_\tau$ is closed under taking direct sums.

(3) $Q_\tau(\_)$ commutes with direct sums.

PROOF: (1) $\Rightarrow$ (2): Let $\{M_i \mid i \in \Omega\}$ be a family of absolutely τ-pure left R-modules. Let $M = \oplus M_i$ and $M' = \Pi M_i$. For each $i \in \Omega$, let $\pi_i: M' \to M_i$ be the canonical projection. Both M and M' are τ-torsion-free and we have a canonical embedding $M \to M'$. We need to show that M is τ-injective.

Let $I \in \mathcal{L}_\tau$ and let $\alpha: I \to M$ be an R-homomorphism. Since each M_i is τ-injective, there exists for each i an R-homomorphism $\beta_i: R \to M_i$ extending $\alpha\pi_i$. These then define an R-homomorphism $\beta: R \to M'$ by $\beta: r \mapsto \langle r\beta_i \rangle$. We want to show that in fact $R\beta \subseteq M$. To do this it suffices to show that $\Lambda = \{i \in \Omega \mid \beta_i \neq 0\}$ is finite. Assume, therefore, that Λ is infinite. In this case we can find a countably infinite subset $\{i_1, i_2, \ldots\}$ of Λ. For each positive integer j, let $I_j = \{r \in I \mid r\beta_{i_k} = 0$ for all $k \geq j\}$. Then $I_1 \subseteq I_2 \subseteq \ldots$ is an ascending chain of left ideals of R and so $\cup I_j = I$ since, for each $r \in I$, we have $r\beta_i = r\alpha\pi_i$ which is nonzero for only finitely many indices $i \in \Omega$. By (1) there exists a positive integer j with $I_j \in \mathcal{L}_\tau$. If $k \geq j$ then $I_j\beta_{i_k} = 0$ and so β_{i_k} induces an R-homomorphism $R/I_j \to M' \in \mathcal{F}_\tau$ which must be the zero map. Therefore, $\beta_{i_k} = 0$ for all $k \geq j$. From this contradiction we deduce that in fact Λ must be finite.

(2) $\Rightarrow$ (3): Let $\{M_i\}$ be a family of left R-modules and let $N = \oplus Q_\tau(M_i)$. Then there exists a canonical

R-homomorphism $\alpha\colon \oplus M_i \to N$ with kernel $\oplus T_\tau(M_i) \in \mathcal{T}_\tau$ and cokernel $N/\oplus M_i \cong \oplus[Q_\tau(M_i)/M_i] \in \mathcal{T}_\tau$. Clearly $N \in \mathcal{F}_\tau$. Also, N is τ-injective by (2) and so $E(N)/N \in \mathcal{F}_\tau$ by Proposition 4.1. Therefore, by Proposition 6.3, $N \cong Q_\tau(\oplus M_i)$.

(3) ⇒ (1): Let $I_1 \subseteq I_2 \subseteq \cdots$ be an ascending chain of left ideals of R with $I = \cup I_j$. For each positive integer j let $\alpha_j\colon I \to I/I_j$ be the canonical R-homomorphism defined by $r \mapsto r + I_j$. Since each $r \in I$ belongs to all but finitely many I_j, the α_j induce an R-homomorphism $\alpha\colon I \to M = \oplus R/I_j$ given by $r \mapsto \langle r + I_j \rangle$. Since $I \in \mathcal{L}_\tau$, there exists an R-homomorphism β making the diagram

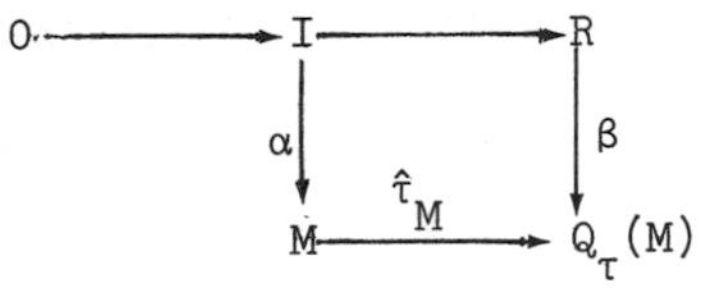

commute. By (3), $Q_\tau(M) \cong \oplus Q_\tau(R/I_j)$ and so there exists an integer k with $R\beta \subseteq \bigoplus_{j=1}^{k} Q_\tau(R/I_j)$.

Pick $h > k$. Then $I\alpha_h\hat{\tau}_M = 0$ and so $I/I_h = I\alpha_h \subseteq T_\tau(R/I_h)$ which implies that I/I_h is τ-torsion. By the exactness of the sequence $0 \to I/I_h \to R/I_h \to R/I \to 0$ we then have $I_h \in \mathcal{L}_\tau$. □

The condition that τ be noetherian does not necessarily imply that $\mathcal{L}_\tau$ satisfies the ascending chain condition. This condition is characterized in the following proposition.

(14.2) PROPOSITION: <u>The following conditions are equivalent for</u> $\tau \in$ R-tors:

(1) $\mathcal{L}_\tau$ <u>satisfies the ascending chain condition</u>.

(2) <u>The class of</u> τ-<u>torsion</u> τ-<u>injective left</u> R-<u>modules is closed under taking direct sums</u>.

PROOF: (1) $\Rightarrow$ (2): Let $\{M_i \mid i \in \Omega\}$ be a family of τ-torsion τ-injective left R-modules and let $M = \oplus M_i$. Then M is clearly τ-torsion. Let $I \in \mathcal{L}_\tau$ and let $\alpha\colon I \to M$ be an R-homomorphism. Since $M \in \mathcal{T}_\tau$, we have $I/\ker(\alpha) \in \mathcal{T}_\tau$. From the exactness of the sequence $0 \to I/\ker(\alpha) \to R/\ker(\alpha) \to R/I \to 0$ we then see that $\ker(\alpha) \in \mathcal{L}_\tau$.

For each $i \in \Omega$ let $\pi_i\colon M \to M_i$ be the canonical projection. We want to show that the set of indices i for which $I\alpha\pi_i \neq 0$ is finite. Assume that this is not so. Then we can pick a countably infinite set $\Lambda = \{i_1, i_2, \ldots\} \subseteq \Omega$ with $I\alpha\pi_{i_j} \neq 0$ for all j. For each positive integer j, set $\Omega_j = (\Omega \setminus \Lambda) \cup \{i_1, \ldots, i_j\}$ and define $I_j = [\oplus \{M_i \mid i \in \Omega_j\}]\alpha^{-1}$. Then $I_1 \subset I_2 \subset \ldots$ is a strictly increasing countably infinite chain of left ideals of R belonging to $\mathcal{L}_\tau$ (since each contains $\ker(\alpha)$).

This contradicts (1). Therefore, $I\alpha\pi_i \neq 0$ only for those i belonging to a finite subset Ω' of Ω. Since $\oplus \{M_i \mid i \in \Omega'\}$ is τ-injective, α can be extended to an R-homomorphism $\beta: R \to M$, proving the τ-injectivity of M.

(2) $\Rightarrow$ (1): If $I \in \mathcal{L}_\tau$ then $E_\tau(R/I)/(R/I)$ is τ-torsion and so, from the exact sequence $0 \to R/I \to E_\tau(R/I) \to E_\tau(R/I)/(R/I) \to 0$, we deduce that $E_\tau(R/I)$ is τ-torsion. Now let $I_1 \subseteq I_2 \subseteq \ldots$ be an ascending chain of left ideals of R belonging to $\mathcal{L}_\tau$ and let $I = \cup I_j$. By (2), $M = \oplus E_\tau(R/I_j)$ is τ-injective and so the R-homomorphism $\alpha: I \to M$, defined by $r \mapsto \langle r + I_j \rangle$, extends to an R-homomorphism $\beta: R \to M$. Since $1 \in R$ maps into an element of M with only finitely many nonzero coordinates, it then follows that there exists a positive integer k such that $I = I_k$. Thus we have (1). □

(14.3) PROPOSITION: <u>Suppose that</u> τ <u>is noetherian and</u> $\mathcal{L}_\tau$ <u>satisfies the ascending chain condition. If</u> $I \in \mathcal{L}_\tau$ <u>and if</u> $\{M_i \mid i \in \Omega\}$ <u>is a family of left</u> R-<u>modules, then for each</u> $\alpha \in \mathrm{Hom}_R(I, \oplus M_i)$ <u>there exists a finite subset</u> Λ <u>of</u> Ω <u>with</u> $I\alpha \subseteq \oplus \{M_i \mid i \in \Lambda\}$.

PROOF: Let $M = \oplus \{M_i \mid i \in \Omega\}$ and for each $i \in \Omega$ let $\pi_i: M \to M_i$ be the canonical projection. Set $\Lambda = \{i \in \Omega \mid I\alpha\pi_i \neq 0\}$. We are done if we can show that Λ

is finite. Assume that it is not. Then Λ contains a countably infinite subset $\Lambda' = \{i_1, i_2, \ldots\}$. For each positive integer k, set $\Lambda_k = (\Lambda \setminus \Lambda') \cup \{i_1, \ldots, i_k\}$ and define $I_k = [\oplus \{M_i \mid i \in \Lambda_k\}]\alpha^{-1}$. Then $\cup I_k = I \in \mathcal{L}_\tau$. Since τ is noetherian, we then have $I_k \in \mathcal{L}_\tau$ for some positive integer k. Then by the ascending chain condition on $\mathcal{L}_\tau$, there exists a $j \geq k$ for which $I_j = I_{j+1} = \ldots$. This contradicts the fact that, by definition, the chain $I_1 \subset I_2 \subset \ldots$ is strictly ascending. Therefore, Λ must be finite. □

We now characterize those $\tau \in R$-tors which are noetherian and for which $\mathcal{L}_\tau$ satisfies the ascending chain condition.

(14.4) PROPOSITION: <u>The following conditions are equivalent for</u> $\tau \in R$-tors:

(1) τ <u>is noetherian and</u> $\mathcal{L}_\tau$ <u>satisfies the ascending chain condition</u>.

(2) <u>Any direct sum of</u> τ-<u>injective left</u> R-<u>modules is</u> τ-<u>injective</u>.

(3) <u>Any direct sum of</u> τ-<u>neat homomorphisms is</u> τ-<u>neat</u>.

PROOF: (1) ⇒ (3): Let $\{\varphi_i : M_i' \to M_i \mid i \in \Omega\}$ be a family of τ-neat R-homomorphisms. Set $M' = \oplus M_i'$,

$M = \oplus M_i$, and $\varphi = \oplus \varphi_i : M' \to M$. To establish (3) it suffices to show that every commutative diagram of the form

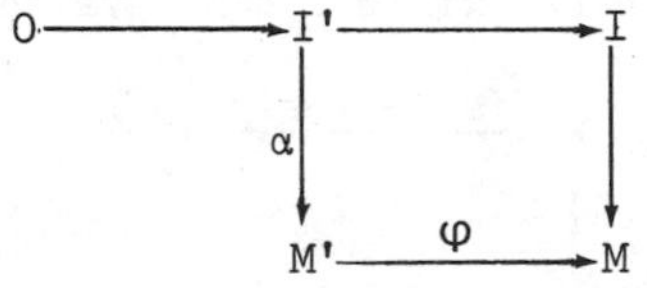

with $I', I \in \mathcal{L}_\tau$ implies the existence of a commutative diagram

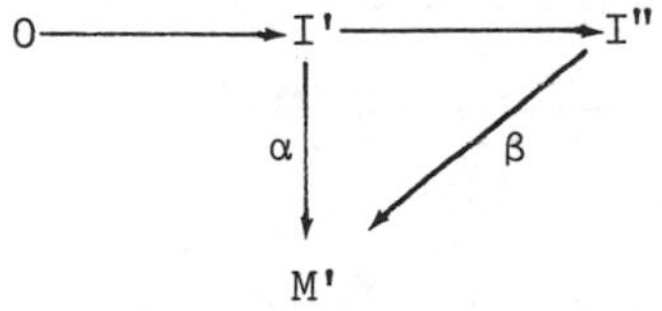

with $I'' \subseteq I$. But by Proposition 14.3 there exists a finite subset Λ of Ω with $I\alpha \subseteq \oplus \{M_i' \mid i \in \Lambda\}$. By Proposition 4.4, $\oplus \{\varphi_i \mid i \in \Lambda\}$ is τ-neat and so such a β does exist.

(3) $\Rightarrow$ (2): This follows from Proposition 4.3.

(2) $\Rightarrow$ (1): This follows from Propositions 14.1 and 14.2. □

We can also consider imposing the ascending chain condition on $\mathcal{C}_\tau$ rather than on $\mathcal{L}_\tau$.

(14.5) PROPOSITION: <u>The following conditions are equivalent for</u> $\tau \in$ R-tors:

(1) $\mathcal{C}_\tau$ <u>satisfies the ascending chain condition</u>.

(2) *If* $I_1 \subset I_2 \subset \dots$ *is a strictly ascending chain of left ideals of* R *with* $I = \cup I_j$, *then there exists a positive integer* k *for which* I_k *is* τ-*dense in* I.

(3) *If* $T_\tau(R) \subset I_1 \subset I_2 \subset \dots$ *is a strictly ascending chain of left ideals of* R *with* $I = \cup I_j$, *then there exists a positive integer* k *for which* I_k *is* τ-*dense in* I.

(4) *Any direct sum of* τ-*torsion-free injective left* R-*modules is injective*.

(5) *If* $E \in \tau$ *then* $E^{(\Omega)}$ *is injective for any index set* Ω.

PROOF: (1) $\Rightarrow$ (2): Let $I_1 \subset I_2 \subset \dots$ be a strictly ascending chain of left ideals of R and let $I = \cup I_j$. For each $j \geq 1$, let H_j be the τ-purification of I_j in R. Then $H_1 \subseteq H_2 \subseteq \dots$ and each $H_j \in \mathcal{C}_\tau$ so there exists a positive integer k with $H_k = H_{k+1} = \dots$. Thus $I \subseteq \cup H_j = H_k$ and so $I/I_k \subseteq H_k/I_k \in \mathcal{J}_\tau$.

(2) $\Rightarrow$ (3): The proof is trivial.

(3) $\Rightarrow$ (4): Let $\{E_j \mid j \in \Omega\}$ be a family of τ-torsion-free injective left R-modules and let $E = \oplus E_j$. To show that E is injective it suffices to show that for every left ideal I of R, every R-homomorphism $\alpha: I \to E$ can be extended to an R-homomorphism $R \to E$.

Let $\alpha: I \to E$ be an R-homomorphism. Since $E \in \mathcal{F}_\tau$, we have $T_\tau(I)\alpha = 0$ and so we can extend α to an R-homomorphism $\alpha': I + T_\tau(R) \to E$ by setting $x\alpha' = 0$ for all $x \in T_\tau(R)$. Thus it suffices to restrict ourselves to those left ideals I of R containing $T_\tau(R)$.

For each left ideal I of R containing $T_\tau(R)$ we define a transfinite sequence H_i of submodules of I as follows:

(a) $H_0 = T_\tau(R)$.

(b) If i is not a limit ordinal and I/H_{i-1} is finitely generated, set $H_i = I$.

(c) If i is not a limit ordinal and I/H_{i-1} is not finitely generated, choose a countably infinite sequence $\langle b_j \rangle$ of elements of I such that $H_{i-1} \subset H_{i-1} + Rb_1 \subset H_{i-1} + Rb_1 + Rb_2 \subset \dots$, and set $H_i = \bigcup_{k=1}^{\infty} [H_{i-1} + \sum_{j=1}^{k} Rb_j]$.

(d) If i is a limit ordinal, set $H_i = \Sigma\{H_j \mid j < i\}$.

Let k(I) be the smallest ordinal i for which $H_i = I$. We shall prove (4) by transfinite induction on k(I). As the base of the induction, we note that if $k(I) = 0$ then $I = T_\tau(R)$, and so the only R-homomorphism $I \to E$ is the zero map, which can trivially be extended to an R-homomorphism $R \to E$.

Assume that $k(I) = i$ and that for any left ideal I' of R with $k(I') < i$, any R-homomorphism $I' \to E$ can be extended to an R-homomorphism $R \to E$. We then have three cases:

<u>Case I</u>: i not a limit ordinal and I/H_{i-1} is finitely generated. If $\alpha: I \to E$ is an R-homomorphism then the restriction of α to H_{i-1} can, by induction, be extended to an R-homomorphism $R \to E$. This implies, in particular, that $H_{i-1}\alpha$ is contained in a direct sum of finitely many of the E_j. Since I/H_{i-1} is finitely generated, $I\alpha$ is also contained in a direct sum of finitely many of the E_j. This direct sum is injective and so α can be extended to an R-homomorphism $R \to E$.

<u>Case II</u>: i is not a limit ordinal and I/H_{i-1} is not finitely generated. By the definition of H_i there then exists a strictly ascending countably infinite chain $H_{i-1} = I_1 \subset I_2 \subset \ldots \subseteq I$ of left ideals of R such that $I = \cup I_j$ and each I_j/H_{i-1} is finitely generated. By (3), there exists a positive integer k such that I_k is τ-dense in I. Since I_k/H_{i-1} is finitely generated, by Case I the restriction of α to I_k can be extended to an R-homomorphism $\beta: R \to E$. Furthermore, $I_k \subseteq \ker(\alpha - \beta_{|I})$ and so $\alpha - \beta_{|I}$ induces an R-homomorphism $I/I_k \to E$, which must be the zero map since I/I_k is τ-torsion. Therfore, $\beta: R \to E$ indeed extends α.

Case III: i is a limit ordinal. Again it suffices to show that if $\alpha: I \to E$ is an R-homomorphism, then $I\alpha$ is contained in a direct sum of finitely many of the E_j. Assume that this is false. Then we can find a countably infinite sequence $\langle i_n \rangle$ of ordinals such that for each $n \geq 1$,

(a) $i_n < k(I)$;

(b) $n < n'$ implies $i_n < i_{n'}$;

(c) $H_{i_n}\alpha$ is not contained in a direct sum of any n of the E_j.

If $\bigcup_{n=1}^{\infty} H_{i_n} \neq I$ then there exists an ordinal $i' < i$ with $\bigcup H_{i_n} \subseteq H_{i'}$, contradicting the induction hypothesis. Therefore, $\bigcup H_{i_n} = I$. We now proceed as in Case II, to show that α can be extended to an R-homomorphism $R \to E$, which is a contradiction.

(4) ⇒ (5): The proof is trivial.

(5) ⇒ (1): Let $E \in \tau$ and let $I_1 \subset I_2 \subset \ldots$ be a countably infinite strictly ascending chain of members of $\mathcal{E}_\tau$. Set $I = \bigcup I_j$. For each j, select $a_j \in I_{j+1} \setminus I_j$. Since I_j is τ-pure in R, there exists an R-homomorphism $\alpha_j: I/I_j \to E$ satisfying $a_j\alpha_j \neq 0$. If $E' = \oplus E_j$ is a countable direct sum of copies of E, then, by (5), E' is injective. But we have an R-homomorphism $\alpha: I \to E'$ defined by $\alpha: a \mapsto \langle (a + I_j)a_j \rangle$ and so, by injectivity, this extends

to an R-homomorphism $\beta\colon R \to E'$. This implies that $I\alpha$ must be contained in a direct sum of finitely many of the E_j, which is a contradiction. Thus (1) is proven. □

(14.6) PROPOSITION: Let $\tau \in R$-tors. A sufficient condition for $\mathcal{C}_\tau$ to satisfy the ascending chain condition is that R_τ be left noetherian.

PROOF: Let $\{E_i\}$ be a family of τ-torsion-free injective left R-modules. Then each E_i is absolutely τ-pure and so is also, by Proposition 6.7, injective as a left R_τ-module. Since R_τ is left noetherian, $\oplus E_i$ is injective as a left R_τ-module and so, by Proposition 6.7, is injective as a left R-module. The result then follows by Proposition 4.5. □

(14.7) PROPOSITION: The following conditions are equivalent for $\tau \in R$-tors:

(1) $\mathcal{C}_\tau$ satisfies the ascending chain condition.

(2) ${}_R(R_\tau)$ satisfies the ascending chain condition on τ-pure left R_τ-ideals.

(3) ${}_R(R_\tau)$ satisfies the ascending chain condition on τ-injective left R_τ-ideals.

PROOF: (1) $\Leftrightarrow$ (2) by Proposition 6.4 and (2) $\Leftrightarrow$ (3) by Proposition 5.4. □

Let $\tau \in R$-tors. A left R-module M is said to be

τ-finitely generated if and only if M has a finitely generated τ-dense submodule.

(14.8) PROPOSITION: Let N be a submodule of a left R-module M. Then

(1) If N and M/N are τ-finitely generated so is M.

(2) If M is τ-finitely generated so is M/N.

(3) M is τ-finitely generated if and only if $M/T_\tau(M)$ is τ-finitely generated.

(4) If M is τ-finitely generated then $Q_\tau(M)$ is τ-finitely generated.

PROOF: (1) Let $\{x_1,\ldots,x_n\} \subseteq N$ and $\{m_1,\ldots,m_k\} \subseteq M$ be selected so that $N' = \Sigma Rx_i$ and $M'/N = \Sigma R(m_j + N)$ are τ-dense submodules of N and M/N respectively. Let $M'' = N' + \Sigma Rm_j \subseteq M'$. Then M'' is a finitely generated submodule of M. Consider the sequence of left R-modules $0 \to N/N' \overset{\alpha}{\to} M/M'' \overset{\beta}{\to} M/M' \to 0$, where α and β are defined respectively by $\alpha: n + N' \mapsto n + M''$ and $\beta: m + M'' \mapsto m + M'$. Then clearly α is a monomorphism and β is an epimorphism. Moreover, $m + M'' \in \ker(\beta) \Leftrightarrow m + N \in M'/N \Leftrightarrow$ there exists an $m' \in M'$ with $m - m' \in N \Leftrightarrow m + M'' \in [M'' + N]/M'' \Leftrightarrow m + M'' \in \mathrm{im}(\alpha)$. Therefore the sequence is exact. Since N/N' and M/M' are τ-torsion by construction, M'' is τ-dense in M.

(2) Let M' be a finitely generated τ-dense submodule of M. Then the R-epimorphism $\nu: M \to M/N$ induces an R-epimorphism $M/M' \to M\nu/M'\nu$. Since M' is τ-dense in M, $M'\nu$ is therefore τ-dense in M/N.

(3) Consider the exact sequence $0 \to T_\tau(M) \to M \to M/T_\tau(M) \to 0$. Then $T_\tau(M)$ is always τ-dense and so the result follows from (1) and (2).

(4) By (3), it suffices to assume that M is τ-torsion-free. In this case we have an exact sequence $0 \to M \to Q_\tau(M) \to Q_\tau(M)/M \to 0$. Since $Q_\tau(M)/M$ is τ-torsion, it is always τ-finitely generated. The result therefore follows by (1). □

(14.9) PROPOSITION: _The following conditions are equivalent for_ $\tau \in R\text{-tors}$:

(1) $\mathcal{C}_\tau$ _satisfies the ascending chain condition_.

(2) _Every left ideal_ I _of_ R _is_ τ-_finitely generated_.

PROOF: (1) ⇒ (2): Let $E \in \tau$. For each finitely generated left ideal I' of R contained in I, let $a(I') = \{x \in E \mid I'x = 0\}$; let $\mathcal{A} = \{a(I') \mid I'$ a finitely generated left ideal of R contained in $I\}$. If $A \in \mathcal{A}$, then $(0:A) \in \mathcal{C}_\tau$ by Proposition 3.9. Therefore, by (1), $\{(0:A) \mid A \in \mathcal{A}\}$ has a maximal element $(0:A_0)$ and so $\mathcal{A}$ has a minimal element $A_0 = a(I_0)$. Let $a \in I \setminus I_0$. Then

$I_1 = I_0 + Ra$ is a finitely generated left ideal of R contained in I and $a(I_1) \subseteq a(I_0)$. By minimality we then have $a(I_1) = a(I_0)$. Since this is true for any $a \in I \setminus I_0$, we in fact have $a(I_0) = \{x \in E \mid Ix = 0\}$.

We claim that I_0 is τ-dense in I. Indeed, assume that it is not. Then there exists a nonzero R-homomorphism $\alpha: I/I_0 \to E$. By the injectivity of E, this can be extended to an R-homomorphism $\beta: R/I_0 \to E$. Let $x_0 = (1 + I_0)\beta$. If $a \in I$ and $(a + I_0)\alpha \neq 0$, then $(a + I_0)\alpha = ax_0$. But if $b \in I_0$, then $bx_0 = (b + I_0)\beta = 0$, so $x_0 \in a(I_0)$ and hence $ax_0 = 0$, which is a contradiction. Therefore, I_0 is τ-dense in I.

(2) ⇒ (1): Let $I_1 \subset I_2 \subset \ldots$ be a strictly ascending chain of left ideals of R with $I = \cup I_j$. By (2), I is τ-finitely generated, and so there exists a finitely generated τ-dense submodule H of I. Then $H \subseteq I_j$ for some j and so I/I_j, being a homomorphic image of I/H, is τ-torsion. Then (1) follows by Proposition 14.5. □

If $\mathcal{C}_\tau$ satisfies the ascending-chain condition then τ is noetherian. Indeed, we have the following result.

(14.10) PROPOSITION: <u>The following conditions are equivalent for</u> $\tau \in R\text{-tors}$:

(1) $\mathcal{C}_\tau$ <u>satisfies the ascending-chain condition</u>.

(2) τ <u>is noetherian and every</u> τ-<u>pure left ideal of</u> R <u>is</u> τ-<u>finitely generated</u>.

PROOF: (1) ⇒ (2): Let $I_1 \subseteq I_2 \subseteq \ldots$ be an ascending chain of left ideals of R the union of which is I. Assume furthermore that $I \in \mathcal{L}_\tau$. If the chain has no strictly ascending subchain then there exists an index k for which $I_k = I$ and so $I_k \in \mathcal{L}_\tau$. Therefore assume that the chain has a strictly ascending subchain. By Proposition 14.5, there exists an index k for which I_k is τ-dense in I. From the exactness of the sequence $0 \to I/I_k \to R/I_k \to R/I \to 0$ it then follows that $I_k \in \mathcal{L}_\tau$. Therefore τ is noetherian. The second part of (2) follows from Proposition 14.9.

(2) ⇒ (1): Let $I_1 \subseteq I_2 \subseteq \ldots$ be an ascending chain of τ-pure left ideals of R and let $I = \cup I_j$. We claim that I is τ-pure in R. Indeed, assume not and let $h \in R \setminus I$ satisfy $h + I \in T_\tau(R/I)$. For each index j, define $H_j = (I_j : h)$. This yields an ascending chain $H_1 \subseteq H_2 \subseteq \ldots$ the union of which is $(I:h)$, which belongs to $\mathcal{L}_\tau$. Since τ is noetherian, this implies that $H_k \in \mathcal{L}_\tau$ for some index k. But $R/H_k \cong [Rh + I_k]/I_k \subseteq R/I_k \in \mathcal{F}_\tau$, a contradiction. Therefore I is τ-pure in R.

By (2), there exists a finitely generated left ideal I' of R which is τ-dense in I. Since I' is finitely generated, $I' \subseteq I_k$ for some index k. Therefore I_k is τ-dense in I_j for all $j \geq k$. But R/I_j is τ-torsion-free for all $j \geq k$ and so we must have $I_j = I_k$ for all $j \geq k$. This proves (1). □

References for Section 14

Beachy [10]; Cunningham [32]; Goldman [63]; Stenström [147]; Teply [152].

15. TORSION THEORIES SATISFYING FINITENESS CONDITIONS

A torsion theory $\tau \in$ R-tors is said to be of *finite type* if and only if $\mathcal{L}_\tau$ has a cofinal subset of finitely generated left ideals. That is to say, τ is of finite type if and only if every $I \in \mathcal{L}_\tau$ is τ-finitely generated. Torsion theories of finite type are clearly noetherian; the converse is not true (see Chapter VII, Example 25). The family of all $\tau \in$ R-tors of finite type will be denoted by R-fin.

(15.1) PROPOSITION: *If $\mathcal{C}_\tau$ satisfies the ascending chain condition then* $\tau \in$ R-fin.

PROOF: This follows from Proposition 14.9. □

(15.2) PROPOSITION: *Let* $\tau \in$ R-fin. *Then any* τ-*pure submodule of a left* R-*module* M *is contained in a maximal* τ-*pure submodule of* M.

PROOF: Let N be a τ-pure submodule of a left R-module M and let $\mathcal{A}$ be the set of all τ-pure submodules of M containing N. Then $\mathcal{A}$ is nonempty since it contains N. Let $\{N_j\}$ be a totally ordered subset of $\mathcal{A}$ and let $N' = \cup N_j$. Pick $x \in M \setminus N'$ and assume that there exists an $I \in \mathcal{L}_\tau$ with $Ix \subseteq N'$. Since τ is of finite type, we can assume without loss of generality that I is finitely generated, say $I = \sum_{i=1}^{k} Ra_i$. Then each $a_i x \in N'$ and so $a_i x \in N_{j(i)}$ for some index $j(i)$. Pick $h \geq \max\{j(1),\ldots,j(n)\}$. Then $Ix \subseteq N_h$, contradicting the fact that N_h is τ-pure in M. We therefore conclude that N' is τ-pure in M. By Zorn's Lemma, $\mathcal{A}$ therefore has a maximal member. □

(15.3) PROPOSITION: <u>The following conditions are equivalent for</u> $\tau \in$ R-tors:

(1) τ <u>is of finite type</u>.

(2) T_τ <u>commutes with direct limits</u>.

(3) <u>A directed union of absolutely</u> τ-<u>pure submodules of a left</u> R-<u>module is absolutely</u> τ-<u>pure</u>.

PROOF: (1) ⇒ (2): Let $\{M_i\}$ be a directed system of left R-modules. The inclusion maps $T_\tau(M_i) \to M_i$ induce an R-monomorphism $\varinjlim T_\tau(M_i) \to \varinjlim M_i$. Since $\mathcal{T}_\tau$ is closed under taking direct sums and homomorphic images, $\varinjlim T_\tau(M_i)$

is τ-torsion and so $\varinjlim T_\tau(M_i) \subseteq T_\tau(\varinjlim M_i)$. Conversely, assume that $x \in T_\tau(\varinjlim M_i)$. Then $(0:x) \in \mathcal{L}_\tau$ and so by (1) there exists a finitely generated left ideal $H \in \mathcal{L}_\tau$ with $Hx = 0$. Since H is finitely generated, there exists a $\langle x_i \rangle \in \oplus M_i$ representing x with $Hx_i = 0$ for all i. Then $x_i \in T_\tau(M_i)$ for all i and so $x \in \varinjlim T_\tau(M_i)$.

(2) ⇒ (1): Let $I \in \mathcal{L}_\tau$ and let $\{I_j\}$ be the set of all finitely generated left ideals of R contained in I. Then canonically $R/I = \varinjlim R/I_j$. Since $R/I \in \mathcal{J}_\tau$, by (2) we have $R/I = \varinjlim T_\tau(R/I_j)$. In particular, there exists an I_k and an $a + I_k \in T_\tau(R/I_k)$ such that $(a + I_k)\nu_k = 1 + I$, where $\nu_k \colon T_\tau(R/I_k) \to R/I$ is the R-homomorphism coming from the direct limit. Then $1 - a \in I$ and so $1 - a \in I_j$ for some j. By taking $\sup\{j,k\}$ if necessary, we can in fact assume that $j = k$ and so $1 - a \in I_k$.

Set $H = (I:a)$. Then $H \in \mathcal{L}_\tau$ and $(Ha/I_k)\nu_k = 0$ and so $Ha \subseteq I_k$. Since $1 - a \in I_k$, we have $H(1 - a) \subseteq I_k$ and so $H \subseteq I_k$. Therefore, $I_k \in \mathcal{L}_\tau$, proving (1).

(1) ⇒ (3): Let $\{N_i\}$ be a directed family of absolutely τ-pure submodules of a left R-module M and let $N = \cup N_i$. Then N is clearly τ-torsion-free and so it remains to show that N is τ-injective. Let $I \in \mathcal{L}_\tau$ and let $\alpha \colon I \to N$ be an R-homomorphism. By (1), I contains a finitely generated subideal H which also belongs to $\mathcal{L}_\tau$. If α' is the restriction of α to H, then $H\alpha' \subseteq N_i$ for

some index i, and so by the τ-injectivity of N_i, α' can be extended to an R-homomorphism $\beta': R \to N_i$. By Proposition 5.1, α and $\beta'|_I$ must be equal and so β' extends α, proving that N is τ-injective.

(3) $\Rightarrow$ (1): Let $I \in \mathcal{L}_\tau$ and let $\{I_j\}$ be the family of finitely generated left ideals of R contained in I. Then I is the directed union of the I_j. By Proposition 6.2 and (3), $Q_\tau(R) \cong Q_\tau(I) = Q_\tau(\cup I_j) \cong \cup Q_\tau(I_j)$. If δ is this isomorphism, then $1\hat{\tau}_R\delta \in Q_\tau(I_k)$ for some index k and so we have a commutative diagram

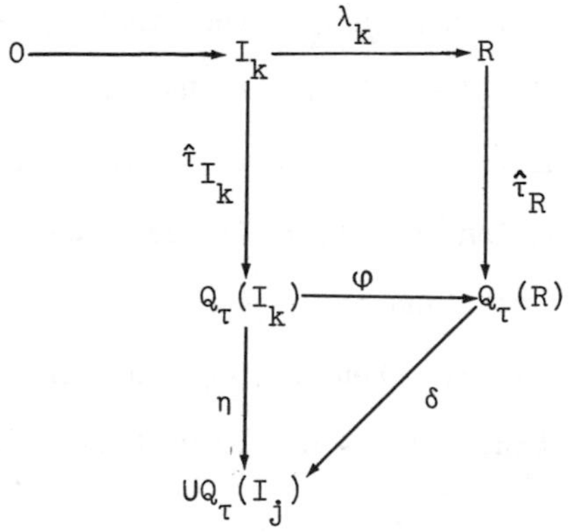

where λ_k and η are the canonical R-monomorphisms and $\varphi = Q_\tau(\lambda_k)$. Since $\varphi\delta = \eta$, φ is a monomorphism. Since $R\hat{\tau}_R\delta \subseteq Q_\tau(I_k)$, we have an induced R-homomorphism $\beta = \hat{\tau}_R\delta\eta^{-1}: R \to Q_\tau(I_k)$. Then $\beta\varphi\delta = \beta\eta = \hat{\tau}_R\delta$ and so $\beta\varphi = \hat{\tau}_R$. Therefore, $\mathrm{im}(\varphi) \cong Q_\tau(I_k)$ is an absolutely τ-pure submodule of $Q_\tau(R)$ containing $R\hat{\tau}_R$ and so must be equal to $Q_\tau(R)$. Thus φ is an isomorphism. By Proposition 6.2, this shows that $I_k \in \mathcal{L}_\tau$. □

The condition that every $I \in \mathcal{L}_\tau$ be finitely generated is stronger than the condition that τ be of finite type.

(15.4) PROPOSITION: The following conditions are equivalent for $\tau \in R\text{-tors}$:

(1) Every $I \in \mathcal{L}_\tau$ is finitely generated.

(2) A directed union of τ-injective submodules of a left R-module M is τ-injective.

(3) If $\{M_i \mid i \in \Omega\}$ and $\{N_i \mid i \in \Omega\}$ are sets of submodules of $M, N \in R\text{-mod}$ respectively, directed by inclusion, and if $\{\varphi_i : M_i \to N_i\}$ is a directed system of τ-neat R-homomorphisms, then $\varphi = \varinjlim \varphi_i : \varinjlim M_i \to \varinjlim N_i$ is a τ-neat homomorphism.

(4) (i) τ is noetherian.

(ii) $\mathcal{L}_\tau$ satisfies the ascending chain condition.

(iii) $\mathcal{L}_\tau$ has a cofinal subset of left ideals of the form $\oplus H_i$ where the H_i are countably generated.

PROOF: (1) ⇒ (2): Let $\{N_i\}$ be a directed set of τ-injective submodules of a left R-module M (directed by inclusion) and let $N = \cup N_i$. If $I \in \mathcal{L}_\tau$ and if $\alpha: I \to N$ is an R-homomorphism, then by (1) $I\alpha \subseteq N_k$ for some index k and so α can be extended to $\beta: R \to N_k$. Hence β followed by the inclusion map $N_k \to N$ extends α, proving that N is τ-injective.

(2) ⇒ (4): This is proven by Propositions 14.4 and 15.3.

(4) ⇒ (1): Let $I \in \mathcal{L}_\tau$. By (4iii) we can find a left ideal H of R contained in I such that $H \in \mathcal{L}_\tau$ and $H = \oplus \{H_i \mid i \in \Omega\}$ where each H_i is nonzero and countably generated. If Ω is infinite then there is a countable subset $\Lambda = \{i_1, i_2, \ldots\}$ of Ω. Then H is the union of the strictly ascending chain of left ideals of the form

$$L_k = \oplus \{H_i \mid i \in (\Omega \setminus \Lambda) \cup \{i_1, \ldots, i_k\}\}.$$

By (4i), $L_n \in \mathcal{L}_\tau$ for some positive integer n. But this contradicts (4ii) since $L_n \subset L_{n+1} \subset \ldots$ is then a strictly increasing chain of left ideals of R belonging to $\mathcal{L}_\tau$.

We conclude, therefore, that Ω is finite and so H is countably generated. If $\{h_1, h_2, \ldots\}$ is a generating set for H, define the left ideals $L_n' = \sum_{i=1}^{n} Rh_i$ for each $n \geq 1$. Then the L_i' form a strictly ascending chain of left ideals of R, the union of which is $H \in \mathcal{L}_\tau$. By (4i), there then exists a positive integer k such that $L_k' \in \mathcal{L}_\tau$. For any sequence $\langle a_i \rangle$ of elements of I, the chain $L_k' \subseteq L_k' + Ra_i \subseteq \ldots$ must terminate by (4ii), and so it follows that for an appropriate choice of the a_i, we have $I = L'_k + \sum_{i=1}^{m} Ra_i$ for some positive integer m. This proves that I is finitely generated.

(3) ⇒ (2): This follows from Proposition 4.3.

(1) ⇒ (3): Consider the commutative diagram

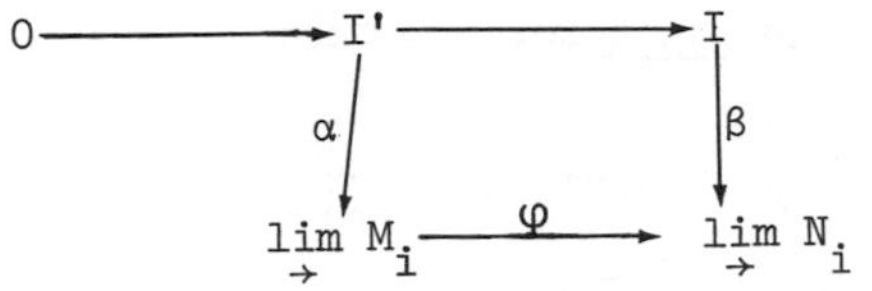

where $I', I \in \mathcal{L}_\tau$. Since both I' and I are finitely generated by (1), there exist $i, j \in \Omega$ such that $I'\alpha \subseteq M_i$ and $I\beta \subseteq N_j$. Choose $k \geq i, j$ in Ω. Then the diagram

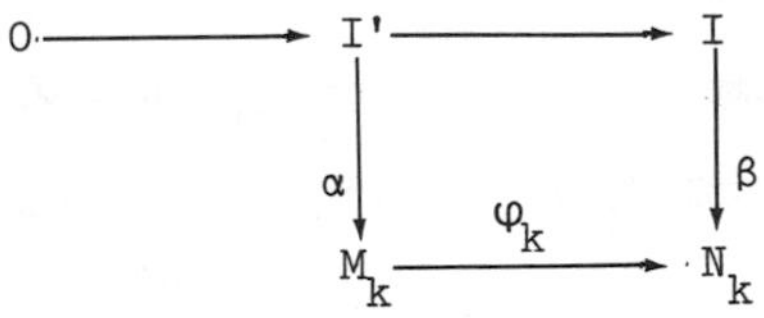

commutes. Since φ_k is τ-neat, there exists a left ideal I'' of R properly containing I' and an R-homomorphism $\beta': I'' \to M_k$ extending α. Consequently β' followed by the canonical R-homomorphism $M_k \to \varinjlim M_i$, is the R-homomorphism needed to show the τ-neatness of φ. □

<u>References for Section 15</u>

Cunningham [32]; Golan and Teply [58]; Goldman [63]; Nastasescu and Popescu [113]; Stenström [147]; Walker and Walker [164].

16. EXACT TORSION THEORIES

Let $\tau \in$ R-tors. The localization functor Q_τ: R-mod $\to$ R-mod is, as we saw in Proposition 6.1, left exact. In general it is not exact (see Chapter VII, Example 26). If $Q_\tau(\_)$ happens to be an exact functor, we say that $\tau \in$ R-tors is an exact torsion theory.

(16.1) PROPOSITION: The following conditions are equivalent for $\tau \in$ R-tors:

(1) τ is exact.

(2) If $M \in \mathcal{E}_\tau$ then $E(M)/M \in \mathcal{E}_\tau$.

(3) If N is a τ-pure submodule of $M \in \mathcal{E}_\tau$ then $M/N \in \mathcal{E}_\tau$.

(4) $\mathrm{Ext}_R^2(N,M) = 0$ for all $N \in \mathcal{T}_\tau$ and $M \in \mathcal{E}_\tau$.

(5) $\mathrm{Ext}_R^1(I,M) = 0$ for all $I \in \mathcal{J}_\tau$ and $M \in \mathcal{E}_\tau$.

PROOF: (1) $\Rightarrow$ (2): Let $M \in \mathcal{E}_\tau$ and consider the exact sequence $0 \to M \to E(M) \to E(M)/M \to 0$. By (1), the sequence $0 \to Q_\tau(M) \to Q_\tau(E(M)) \to Q_\tau(E(M)/M) \to 0$ is exact. But $M = Q_\tau(M)$ since M is absolutely τ-pure and $Q_\tau(E(M)) = E(M)$ since $E(M)$ is also absolutely τ-pure. Therefore, we must have $E(M)/M = Q_\tau(E(M)/M)$, proving that $E(M)/M$ is absolutely τ-pure.

(2) $\Rightarrow$ (3): Let N be a τ-pure submodule of $M \in \mathcal{E}_\tau$. We then have the exact sequence $0 \to M/N \to E(M)/N \to E(M)/M \to 0$ where M/N is τ-torsion-free by the τ-purity of N, and

$E(M)/M$ is τ-torsion-free by Proposition 4.1. Therefore, $E(M)/N$ is τ-torsion-free. Moreover, $E(N)$ is a direct summand of $E(M)$, say $E(M) = E(N) \oplus M'$. Therefore, $E(M)/N \cong E(N)/N \oplus M'$, where the first summand is τ-injective by (2) and the second summand is injective since it is a direct summand of $E(M)$. Therefore, $E(M)/N$ is τ-injective and so absolutely τ-pure. Since $[E(M)/N]/[M/N] \cong E(M)/M \in \mathcal{F}_\tau$, then M/N is absolutely τ-pure by Proposition 5.4.

(3) ⇒ (1): Let $0 \to M' \to M \xrightarrow{\alpha} M'' \to 0$ be an exact sequence and let $N = \mathrm{im}(Q_\tau(\alpha)) \subseteq Q_\tau(M'')$. By (3), N is absolutely τ-pure. Since N clearly contains M'', then by uniqueness we must have $N = Q_\tau(M'')$, proving exactness.

(2) ⇒ (4): Let N be a τ-torsion module and let $M \in \mathcal{E}_\tau$. Then $E(M) \in \mathcal{E}_\tau$ and by (3) so is $M' = E(M)/M$. We first note that from the exact sequence $0 \to M \to E(M) \to M' \to 0$ we obtain the exact sequence $\mathrm{Ext}^1_R(N,E(M)) \to \mathrm{Ext}^1_R(N,M') \to \mathrm{Ext}^2_R(N,M) \to \mathrm{Ext}^2_R(N,E(M))$, the end terms of which are equal to 0 since $E(M)$ is injective. Thus $\mathrm{Ext}^1_R(N,M') \cong \mathrm{Ext}^2_R(N,M)$. But $\mathrm{Ext}^1_R(N,M') = 0$ by Proposition 4.1, proving (4).

(4) ⇒ (5): If $I \in \mathcal{L}_\tau$ then R/I is τ-torsion, and from the exact sequence $0 \to I \to R \to R/I \to 0$ we obtain the isomorphism $\mathrm{Ext}^1_R(I,M) \cong \mathrm{Ext}^2_R(R/I,M)$, whence (5) follows directly from (4).

(5) ⇒ (2): Let $M \in \mathcal{E}_\tau$ and let $I \in \mathcal{L}_\tau$. Then from the exact sequence $0 \to M \to E(M) \to E(M)/M \to 0$ we derive the exact sequence $0 \to \mathrm{Hom}_R(I,M) \to \mathrm{Hom}_R(I,E(M)) \to \mathrm{Hom}_R(I,E(M)/M) \to \mathrm{Ext}^1_R(I,M)$ the right-hand term of which equals 0 by (5).

By Proposition 4.1, $E(M)/M$ is τ-torsion-free. If $\alpha \in \mathrm{Hom}_R(I,E(M)/M)$, then by the preceding paragraph there exists an R-homomorphism $\beta: I \to E(M)$ making the diagram

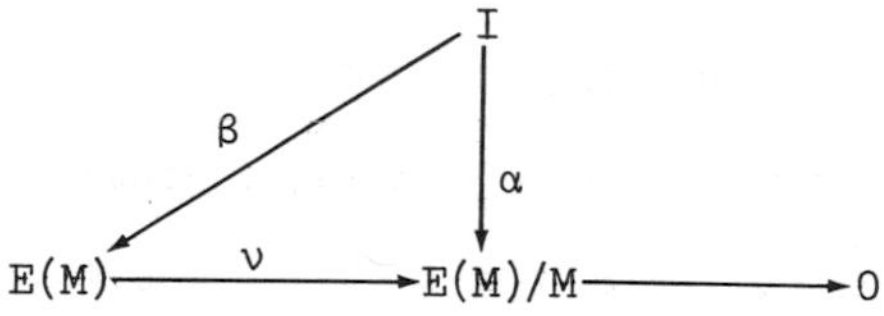

commute. By the injectivity of $E(M)$ the R-homomorphism β can be extended to an R-homomorphism $\varphi: R \to E(M)$. Then $\varphi\nu$ extends α and so $E(M)/M$ is τ-injective. Thus $E(M)/M \in \mathcal{E}_\tau$. □

By Propositions 16.1(3) and 5.6, we immediately see that the exactness of the functor $T_\tau(\_)$ implies the exactness of τ.

Let $\tau \in$ R-tors. We say that a left R-module M is τ-<u>projective</u> if and only if every diagram of the form

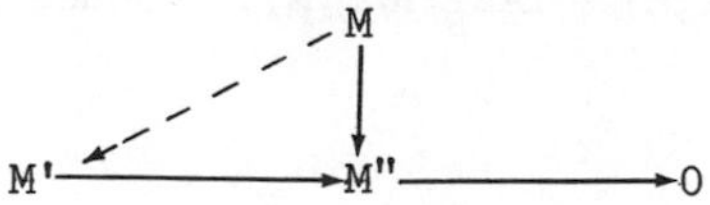

with $M' \in \mathcal{E}_\tau$ and $M'' \in \mathcal{F}_\tau$ can be completed commutatively. For example, if M is a projective left R-module, then $M/T_\tau(M)$ is clearly τ-projective.

(16.2) PROPOSITION: *Let* $\tau \in$ R-tors. *Then the following conditions are equivalent for a left R-module* M:

(1) M *is* τ-*projective*.

(2) *If* $M', M'' \in \mathcal{F}_\tau$, *if* $\alpha: M' \to M''$ *is an R-epimorphism, and if* $\beta: M \to M''$ *is an R-homomorphism, then there exists a* τ-*dense submodule* N *of* M *and an R-homomorphism* $\beta': N \to M'$ *making the following diagram commute*.

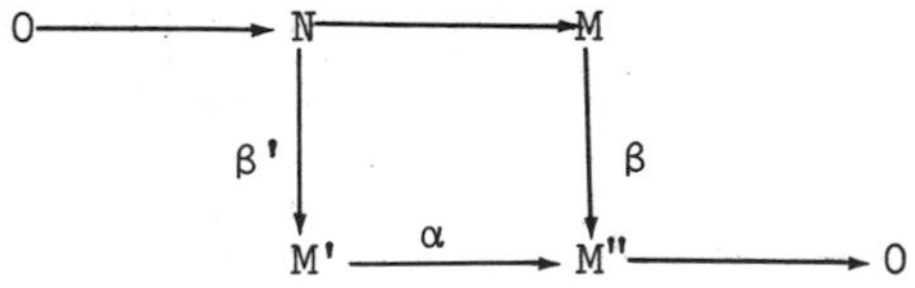

PROOF: (1) ⇒ (2): Given an R-epimorphism $\alpha: M' \to M''$ with $M', M'' \in \mathcal{F}_\tau$, the R-homomorphism $\alpha' = Q_\tau(\alpha)$ makes the diagram

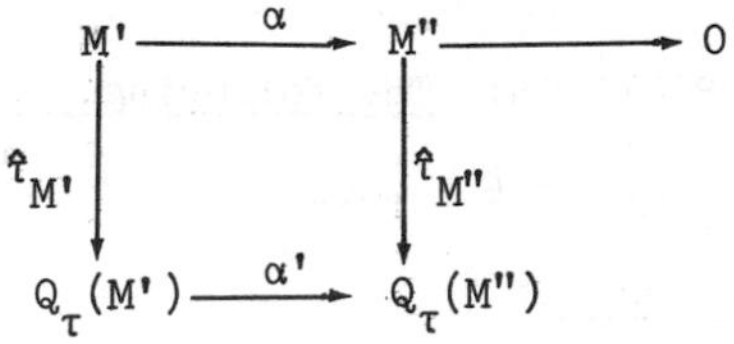

commute and so $\mathrm{im}(\hat{\tau}_{M''}) \subseteq \mathrm{im}(\alpha')$. By (1), there then exists an R-homomorphism φ making the diagram

$$\begin{array}{ccccc} M & \xrightarrow{\beta} & M'' & & \\ \varphi\downarrow & & \downarrow\hat{\tau}_{M''} & & \\ Q_\tau(M') & \xrightarrow{\alpha'} & \mathrm{im}(\alpha') & \longrightarrow & 0 \end{array}$$

commute, since $Q_\tau(M') \in \mathcal{E}_\tau$ and $\mathrm{im}(\alpha') \in \mathcal{F}_\tau$. Set $N = (M'\hat{\tau}_{M'})\varphi^{-1}$ and let β' be the restriction of φ to N.

(2) ⇒ (1): Consider an R-epimorphism $\alpha: M' \to M''$ and an R-homomorphism $\beta: M \to M''$, where $M' \in \mathcal{E}_\tau$ and $M'' \in \mathcal{F}_\tau$. By (2), there exists a τ-dense submodule of M and an R-homomorphism β' making the diagram

$$\begin{array}{ccccccc} 0 & \longrightarrow & N & \longrightarrow & M & & \\ & & \beta\downarrow & & \downarrow\beta & & \\ & & M' & \xrightarrow{\alpha} & M'' & \longrightarrow & 0 \end{array}$$

commute. Since M' is τ-injective and N is τ-dense, β' extends to an R-homomorphism $\varphi: M \to M'$. Moreover, $N(\varphi\alpha - \beta) = 0$ and so $\varphi\alpha - \beta$ induces an R-homomorphism $M/N \to M''$ which must be the zero map since $M/N \in \mathcal{T}_\tau$ and $M'' \in \mathcal{F}_\tau$. Therefore, $\varphi\alpha = \beta$, proving (1). □

(16.3) PROPOSITION: <u>The following conditions are equivalent for</u> $\tau \in$ R-tors:

(1) τ <u>is exact</u>.

(2) <u>Every</u> $I \in \mathcal{L}_\tau$ <u>is</u> τ-<u>projective</u>.

PROOF: (1) ⇒ (2): Let $I \in \mathcal{L}_\tau$, let $\alpha: M' \to M''$ be an R-epimorphism with $M' \in \mathcal{E}_\tau$ and $M'' \in \mathcal{F}_\tau$, and let $\beta: I \to M''$ be an R-homomorphism. By Proposition 16.1, M'' is absolutely τ-pure and so β can be extended to an R-homomorphism $\varphi: R \to M''$. Since R is projective, there exists an R-homomorphism $\psi: R \to M'$ satisfying $\varphi = \psi\alpha$. If ψ' is the restriction of ψ to I, then $\psi'\alpha = \beta$, proving that I is τ-projective.

(2) ⇒ (1): If $M \in \mathcal{E}_\tau$ then $E(M) \in \mathcal{E}_\tau$, and so we have the exact sequence $0 \to M \to E(M) \to E(M)/M \to 0$. If $I \in \mathcal{L}_\tau$, we then derive the exact sequence

$$\mathrm{Hom}_R(I,E(M)) \xrightarrow{\theta} \mathrm{Hom}_R(I,E(M)/M) \to \mathrm{Ext}^1_R(I,M) \to \mathrm{Ext}^1_R(I,E(M)).$$

Since $E(M)$ is injective, the right-hand end is 0. Furthermore, by (2), θ is surjection and so $\mathrm{Ext}^1_R(I,M) = 0$. Then (1) follows by Proposition 16.1. □

(16.4) COROLLARY: <u>If</u> R <u>is left hereditary then every</u> $\tau \in R\text{-tors}$ <u>is exact</u>.

PROOF: If R is left hereditary then every left ideal of R is projective and so it is surely τ-projective. □

In Proposition 6.6 we saw that if M is an absolutely τ-pure left R-module, then M is also an R_τ-module in a manner that naturally extends its R-module structure. With the aid of the notion of a τ-projective module, we can characterize all left R-modules having this property.

(16.5) PROPOSITION: Let $\tau \in$ R-tors. Then the following conditions are equivalent for a τ-torsion-free left R-module M:

(1) M is the homomorphic image of a direct sum of copies of $Q_\tau(R)$.

(2) If N' is a finitely generated τ-projective τ-dense submodule of a left R-module N, then every R-homomorphism from N' to M extends to an R-homomorphism from N to M.

(3) M has an R_τ-module structure which naturally extends its structure as a left R-module.

PROOF: (1) $\Rightarrow$ (2): Let N' be a finitely generated τ-projective τ-dense submodule of a left R-module N and let α: N' $\to$ M be an R-homomorphism. By (1), there exists an R-epimorphism $\beta: Q_\tau(R)^{(\Omega)} \to M$ for some index set Ω. Since N' is finitely generated, $N'\alpha \subseteq Q_\tau(R)^n\beta$ for some positive integer n. Since $Q_\tau(R)^n$ is absolutely τ-pure and N' is τ-projective, there then exists an R-homomorphism α' making the diagram

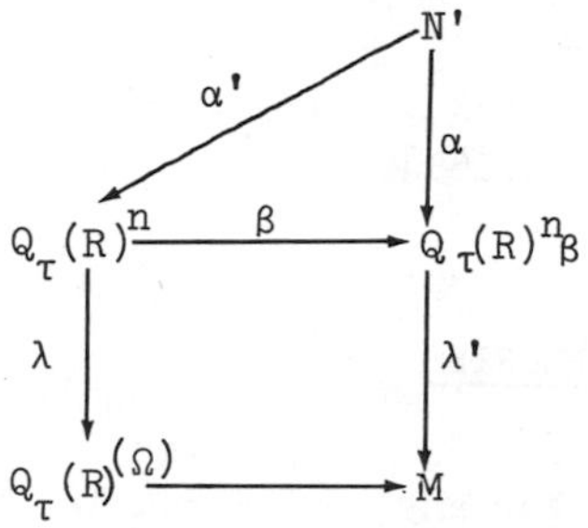

commute, where λ and λ' are inclusion maps. Moreover, by the τ-injectivity of $Q_\tau(R)^n$ and the fact that N' is τ-dense in M, there exists an R-homomorphism $\varphi\colon N \to Q_\tau(R)^n$ extending α'. Therefore, $\varphi\beta\lambda'\colon N \to M$ extends α, proving (2).

(2) ⇒ (3): Since R is projective as a left R-module over itself, $R/T_\tau(R)$ is τ-projective. Since M is τ-torsion-free, for each $m \in M$ there exists an R-homomorphism $R \to M$ defined by $r \mapsto rm$ and this induces an R-homomorphism $R/T_\tau(R) \to M$. By (2), this homomorphism extends to an R-homomorphism $Q_\tau(R) \to M$ that is unique since M is τ-torsion-free. Since $Q_\tau(R) \cong R_\tau$, this defines an R_τ-module structure on M that naturally extends the R-module structure of M.

(3) ⇒ (1): As a left R_τ-module, M is the homomorphic image of the free R_τ-module $Q_\tau(R)^{(\Omega)}$ for some index set Ω. Viewing these as left R-modules, we obtain (1). □

References for Section 16

Beachy [10, 12]; Chamard [26]; Goldman [63]; Hacque [66]; Lambek [92, 93]; Winton [165].

17. PERFECT TORSION THEORIES

Let $\tau \in R\text{-tors}$. For any left R-module M, the R-homomorphism $\hat{\tau}_M: M \to Q_\tau(M)$ can be factored as

$$M \xrightarrow{\zeta_M} R_\tau \otimes_R M \xrightarrow{\eta_M} Q_\tau(M)$$

where $\zeta_M: m \mapsto 1 \otimes m$ and $\eta_M: \Sigma(s_i \otimes m_i) \mapsto \Sigma s_i(m_i\hat{\tau}_M)$. In general, η_M is not an R-isomorphism. If, however, it is true that η_M is an R-isomorphism for every left R-module M, we say that τ is *perfect*. Perfect torsion theories can be characterized in many ways, as the following result shows.

(17.1) PROPOSITION: *The following conditions are equivalent for* $\tau \in R\text{-tors}$:

(1) τ *is perfect*.

(2) $\ker(\zeta_M) = T_\tau(M)$ *for every left R-module* M.

(3) $R_\tau\hat{\tau}(I) = R_\tau$ *for every* $I \in \mathcal{L}_\tau$.

(4) *Every left* R_τ*-module is absolutely* τ*-pure as a left R-module*.

(5) *Every left* R_τ*-module is* τ*-torsion-free as a left R-module*.

(6) There exists a cogenerator of R_τ-mod which is τ-torsion-free as a left R-module.

(7) Every simple left R_τ-module is τ-torsion-free as a left R-module.

(8) τ is exact and noetherian.

(9) τ is exact and of finite type.

PROOF: (1) ⇒ (2): By (1), $\ker(\zeta_M) = \ker(\hat{\tau}_M) = T_\tau(M)$ for every left R-module M.

(2) ⇒ (3): If $I \in \mathcal{L}_\tau$, then by (2), $\zeta_{R/I} = 0$. But $(R/I)\zeta_{R/I} = R_\tau \otimes_R R/I \cong R_\tau/R_\tau\hat{\tau}(I)$ and so $R_\tau = R_\tau\hat{\tau}(I)$.

(3) ⇒ (4): Let N be a left R_τ-module. If $x \in T_\tau({}_RN)$ then there exists an $I \in \mathcal{L}_\tau$ with $0 = Ix = \hat{\tau}(I)x$. But then $x \in R_\tau x = R_\tau\hat{\tau}(I)x = 0$, and so $x = 0$. Therefore, ${}_RN$ is τ-torsion-free.

Now let $I \in \mathcal{L}_\tau$ and let $\alpha: I \to N$ be an R-homomorphism. Since ${}_RN$ is τ-torsion-free, $T_\tau(I)\alpha = 0$ and so α induces an R-homomorphism $\alpha': {}_R\hat{\tau}(I) \to N$. By (3), $1 = \Sigma s_i\hat{\tau}(a_i)$ for some $s_i \in R_\tau$ and $a_i \in I$. Set $H = I \cap [\cap_i (\hat{\tau}(R): s_i)] \in \mathcal{L}_\tau$ and let $x = \Sigma s_i(a_i\alpha) \in N$. Define the R-homomorphism $\beta: R \to N$ by $r \mapsto rx$. If $h \in H$ then $h\beta = \Sigma hs_i(a_i\alpha) = \Sigma h_is_i[\hat{\tau}(a_i)\alpha'] = \Sigma[hs_i\hat{\tau}(a_i)]\alpha' = (h \cdot 1)\alpha' = h\alpha$ and so the restriction of β to H equals the restriction of α to H. Therefore, $\beta_{|I} - \alpha$ induces an R-homomorphism $I/H \to N$ which must be the zero map since $I/H \in \mathcal{J}_\tau$. Therefore, the restriction of β to I agrees with α, proving that N is τ-injective as a left R-module and hence absolutely τ-pure.

(4) ⇒ (5): The proof is trivial.

(5) ⇒ (3): For any $I \in \mathcal{L}_\tau$ we have $\hat{\tau}(I) \subseteq \hat{\tau}(R) \subseteq R_\tau$. By the definition of R_τ, $\hat{\tau}(R)$ is τ-dense in R_τ. Also, $\hat{\tau}(R)/\hat{\tau}(I) \cong R/[I + T_\tau(R)]$ and so is τ-torsion. Therefore, by the exactness of the sequence of left R-modules $0 \to \hat{\tau}(R)/\hat{\tau}(I) \to R_\tau/\hat{\tau}(I) \to R_\tau/\hat{\tau}(R) \to 0$, we see that $\hat{\tau}(I)$ is τ-dense in R_τ. Since $\hat{\tau}(I) \subseteq R_\tau\hat{\tau}(I)$, we have $R_\tau/R_\tau\hat{\tau}(I) \in \mathcal{J}_\tau$. But by (5), $R_\tau/R_\tau\hat{\tau}(I) \in \mathcal{F}_\tau$, whence $R_\tau = R_\tau\hat{\tau}(I)$.

(5) ⇒ (7): The proof is trivial.

(7) ⇒ (6): For each $N \in R_\tau$-simp, let $E'(N)$ be the injective hull of N in R_τ-mod. Then $\Pi E'(N)$ is a cogenerator of R_τ-mod. By Corollary 6.8, each $E'(N)$ is also the injective hull of ${}_RN$ and so $\Pi E'(N)$ is τ-torsion-free as a left R-module. Thus we have (6).

(6) ⇒ (5): Let N be a cogenerator of R_τ-mod that is τ-torsion-free as a left R-module. If N' is any other left R_τ-module, then there exists an R_τ-monomorphism $N' \to N^\Omega$ for some index set Ω. Since N is τ-torsion-free, so is N^Ω and hence so is N'.

(4) ⇒ (1): For any left R-module M, $R_\tau \otimes_R M$ is τ-torsion-free as a left R-module by (4). Therefore,

$T_\tau(M) \subseteq \ker(\zeta_M)$. But then $\ker(\zeta_M) \subseteq \ker(\hat{\tau}_M) = T_\tau(M)$ and so $\ker(\zeta_M) = T_\tau(M) \in \mathcal{J}_\tau$. Moreover, it is trivial to show that $\mathrm{coker}(\zeta_M) \in \mathcal{J}_\tau$. Since $R_\tau \otimes_R M$ is absolutely τ-pure, $E(R_\tau \otimes_R M)/(R_\tau \otimes_R M)$ is τ-torsion-free by Proposition 4.1. Therefore, η_M is an isomorphism by Proposition 6.3.

(1) ⇒ (9): The proof is immediate using Proposition 15.3(3) since tensor products preserve direct limits.

(9) ⇒ (8): The proof is trivial.

(8) ⇒ (3): Let $I \in \mathcal{L}_\tau$. Then $R_\tau \hat{\tau}(I)$ is a left R_τ-module and so $R_\tau \hat{\tau}(I)$ is the homomorphic image of $R_\tau^{(\Omega)}$ for some index set Ω. By (8), $R_\tau^{(\Omega)} \cong Q_\tau(R^{(\Omega)})$ and so is absolutely τ-pure. Moreover, $R_\tau \hat{\tau}(I)$ is a submodule of R_τ and so, as a left R-module, is τ-torsion-free. By the exactness of τ and by Proposition 16.1, this implies that $R_\tau \hat{\tau}(I)$ is absolutely τ-pure as a left R-module. Moreover, in the proof of (5) ⇒ (3) we saw that $R_\tau \hat{\tau}(I)$ is τ-dense in R_τ and so, by τ-injectivity, the identity map $R_\tau \hat{\tau}(I) \to R_\tau \hat{\tau}(I)$ can be extended to an R-homomorphism $R_\tau \to R_\tau \hat{\tau}(I)$. Thus $R_\tau \hat{\tau}(I)$ is a direct summand of R_τ. Since R_τ is τ-torsion-free as a left R-module, we must then in fact have $R_\tau \hat{\tau}(I) = R_\tau$. □

(17.2) COROLLARY: <u>If</u> $T_\tau(\_)$ <u>is exact then</u> $\tau \in R$-tors <u>is perfect</u>.

PROOF: Let $\tau \in$ R-tors with $T_\tau(\_)$ exact. If $I \in \mathcal{L}_\tau$, then by Proposition 5.5, $R = I + T_\tau(R)$ and so there exist $a \in I$ and $b \in T_\tau(R)$ such that $1 = a + b$. Then $1 = \hat{\tau}(1) = \hat{\tau}(a) + \hat{\tau}(b) = \hat{\tau}(a)$ and so $1 \in \hat{\tau}(I)$. Hence $R_\tau \hat{\tau}(I) = R_\tau$, proving that τ is perfect by Proposition 17.1. □

(17.3) COROLLARY: _If_ $\tau \in$ R-tors _is faithful and if every proper left ideal of_ R_τ _has a nonzero right annihilator, then_ τ _is perfect_.

PROOF: Since τ is faithful, R is a submodule of R_τ. If $I \in \mathcal{L}_\tau$ and $R_\tau I$ is a proper left ideal of R_τ, then by hypothesis there exists $0 \neq s \in R_\tau$ with $R_\tau Is = 0$. But then $Is = 0$, contradicting the fact that R_τ is τ-torsion-free. Therefore, we must have $R_\tau I = R_\tau$ whence τ is perfect by Proposition 17.1. □

The family of all perfect $\tau \in$ R-tors will be denoted by R-perf.

(17.4) PROPOSITION: R-perf _is closed under taking arbitrary joins_.

PROOF: Let $U \subseteq$ R-perf and let $M \in R_{\vee U}$-mod. Then $M \in R_\tau$-mod for all $\tau \in U$ by Proposition 7.3. Therefore, by Proposition 17.1, $M \in \mathcal{F}_\tau$ for all $\tau \in U$ whence $M \in \cap \mathcal{F}_\tau = \mathcal{F}_{\vee U}$. By Proposition 17.1, this implies that $\vee U$

is perfect. □

(17.5) PROPOSITION: *If* $\tau \in$ R-perf *then there exists a bijective correspondence between the left ideals of* R_τ *and the* τ*-pure left ideals of* R *given by* $H \mapsto \hat{\tau}^{-1}(H)$ *and* $I \mapsto Q_\tau(I)$.

PROOF: By Proposition 17.1, every left ideal of R_τ is τ-pure in R_τ when considered as a left R-submodule. The result then follows by Proposition 6.4. □

(17.6) COROLLARY: *The following conditions are equivalent for an exact* $\tau \in$ R-tors:

(1) $\mathcal{C}_\tau$ *satisfies the ascending-chain condition*.

(2) R_τ *is left noetherian*.

PROOF: (2) ⇒ (1): This follows from Proposition 14.6.

(1) ⇒ (2): By Proposition 15.1, $\tau \in$ R-fin and so, by Proposition 17.1, τ is perfect. The result then follows from Proposition 17.5. □

We now turn to another method of characterizing perfect torsion theories by considering the properties of the ring homomorphism $\hat{\tau}$.

(17.7) PROPOSITION: *The following conditions are equivalent for a ring homomorphism* $\gamma: R \to S$:

(1) γ <u>is an epimorphism in the cateogory of rings</u>.

(2) <u>If</u> $M \in$ S-mod-S <u>and if</u> $m \in M$ <u>satisfies</u> $\gamma(r)m = m\gamma(r)$ <u>for every</u> $r \in R$, <u>then</u> $sm = ms$ <u>for every</u> $s \in S$.

(3) <u>The functor</u> γ_*: S-mod $\to$ R-mod <u>is full</u>.

(4) <u>The adjunction transformation</u> $\gamma^*\gamma_* \to$ <u>identity on</u> S-mod <u>is a natural equivalence</u>.

(5) <u>The S-homomorphism</u> $\theta: S \otimes_R S \to S$ <u>defined by</u> $\Sigma(s_i \otimes s_i') \mapsto \Sigma s_i s_i'$ <u>is an isomorphism</u>.

PROOF: (1) $\Rightarrow$ (2): Let $M \in$ S-mod-S and let A be the ring of matrices of the form $\begin{bmatrix} s & m \\ 0 & s' \end{bmatrix}$ where $m \in M$ and $s, s' \in S$. Then we have a canonical ring homomorphism $\delta: S \to A$ defined by $\delta: s \mapsto \begin{bmatrix} s & 0 \\ 0 & s \end{bmatrix}$. Furthermore, for each $m \in M$ we also have a ring homomorphism $\delta_m: S \to A$ defined by $\delta_m: s \mapsto \begin{bmatrix} s & sm-ms \\ 0 & s \end{bmatrix}$. If $\gamma(r)m = m\gamma(r)$ for every $r \in R$, then $\delta\gamma = \delta_m\gamma$ and so, by (1), $\delta = \delta_m$, whence $sm = ms$ for every $s \in S$.

(2) $\Rightarrow$ (3): Let N, N' be left S-modules. Then $\mathrm{Hom}_{\mathbb{Z}}(N,N') \in$ S-mod-S under the definitions $\alpha s: n \mapsto (sn)\alpha$ and $s\alpha: n \mapsto s(n\alpha)$ for all $s \in S$, $n \in N$, and $\alpha \in \mathrm{Hom}_{\mathbb{Z}}(N,N')$. If $\alpha \in \mathrm{Hom}_R(N,N') \subseteq \mathrm{Hom}_{\mathbb{Z}}(N,N')$, then $(rn)\alpha - r(n\alpha) = 0$ for all $r \in R$ and $n \in N$ and so $\gamma(r)\alpha = \alpha\gamma(r)$ for all $r \in R$. By (2), $s\alpha = \alpha s$ for all $s \in S$ and so $\alpha \in \mathrm{Hom}_S(N,N')$.

(3) ⇒ (4): We have to show that for every left S-module N the S-homomorphism $\alpha: S \otimes_R N \to N$ defined by $\Sigma(s_i \otimes x_i) \mapsto \Sigma s_i x_i$ is an isomorphism. The map $\beta: N \to S \otimes_R N$ defined by $n \mapsto 1 \otimes n$ is clearly an R-homomorphism, and so by (3) it is an S-homomorphism. It therefore must be the inverse of α, proving that α is an isomorphism.

(4) ⇒ (5): The proof is trivial.

(5) ⇒ (1): Let $\delta_1, \delta_2: S \to S'$ be ring homomorphisms satisfying $\delta_1\gamma = \delta_2\gamma$. Then we have a well-defined homomorphism $\delta: S \otimes_R S \to S'$ given by $\delta: \Sigma(s_i \otimes s_i') \mapsto \Sigma\delta_1(s_i)\delta_2(s_i')$. Since the map θ defined in (5) is an isomorphism, for each $s \in S$ we have $s \otimes 1 = \theta^{-1}(s) = 1 \otimes s$, and so in particular for each $s \in S$, we have $\delta_1(s) = \delta(s \otimes 1) = \delta(1 \otimes s) = \delta_2(s)$. Therefore, $\delta_1 = \delta_2$ and so γ is a ring epimorphism. □

We say that a ring epimorphism $\gamma: R \to S$ is left flat if and only if S is flat as a right R-module.

(17.8) PROPOSITION: The following conditions are equivalent for a ring homomorphism $\gamma: R \to S$:

(1) γ is a left flat ring epimorphism.

(2) γ is a ring epimorphism and every injective left S-module is injective as a left R-module.

(3) There exists a $\tau \in$ R-perf satisfying

$\mathcal{L}_\tau = \{_R I \subseteq R \mid S\gamma(I) = S\}$ <u>and with</u> $Q_\tau(\underline{\mathcal{L}})$ <u>naturally equivalent to</u> $\gamma^* = S \otimes_R \_$.

PROOF: (1) ⇒ (2): Let N be an injective left S-module. Since S_R is flat, the functor $S \otimes_R \_ : R\text{-mod} \to S\text{-mod}$ is exact. Let $0 \to M' \to M \to M'' \to 0$ be a short exact sequence of left R-modules. Then $0 \to S \otimes_R M' \to S \otimes_R M \to S \otimes_R M'' \to 0$ is exact and so, by the injectivity of N, the sequence $0 \to \mathrm{Hom}_S(S \otimes_R M', N) \to \mathrm{Hom}_S(S \otimes_R M, N) \to \mathrm{Hom}_S(S \otimes_R M'', N) \to 0$ is exact. Applying the canonical natural transformation, we obtain the exact sequence of abelian groups $0 \to \mathrm{Hom}_R(M',N) \to \mathrm{Hom}_R(M,N) \to \mathrm{Hom}_R(M'',N) \to 0$, proving that N is injective as a left R-module.

(2) ⇒ (1): Let N' be a submodule of a left R-module M and let N be the injective hull of $S \otimes_R M'$ in S-mod. Then by (2), N is injective as a left R-module. Therefore, if $\lambda: M' \to M$ is the inclusion map and if $\beta: M' \to S \otimes_R M'$ is the R-homomorphism defined by $x \mapsto 1 \otimes x$, there exists an R-homomorphism ψ making the diagram

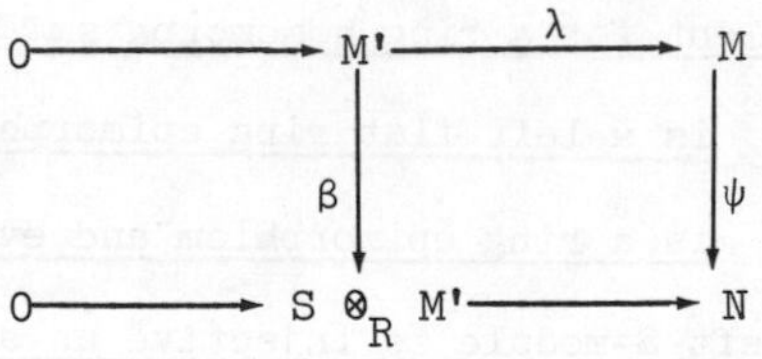

commute. Applying the functor $S \otimes_R \_$ to this diagram and recalling that since γ is a ring epimorphism we have $S \otimes_R W = W$ for every left S-module W (see Proposition 17.7), we obtain the commutative diagram

$$\begin{array}{ccccc} & & S \otimes_R M' & \xrightarrow{1 \otimes \lambda} & S \otimes_R M \\ & & \Big\downarrow = & & \Big\downarrow 1 \otimes \psi \\ 0 & \longrightarrow & S \otimes_R M' & \longrightarrow & N \end{array}$$

which proves that $1 \otimes \lambda$ is an S-monomorphism. Therefore, S_R is flat.

(1) ⇒ (3): By Proposition 17.7, the functor γ_*: S-mod → R-mod is full and $\gamma^*: S \otimes_R \_$ is its left adjoint. Since S_R is flat, γ^* is exact. Let $\mathcal{A} = \{M \in \text{R-mod} \mid \gamma^*(M) = 0\}$. Since γ^* has a right adjoint, it commutes with direct sums and so $\mathcal{A}$ is closed under taking direct sums. Since γ^* is exact, $\mathcal{A}$ satisfies the other conditions of Proposition 1.6(2) and so $\mathcal{A} = \mathcal{J}_\tau$ for some $\tau \in$ R-tors. Moreover, $\mathcal{L}_\tau = \{{}_RI \subseteq R \mid S \otimes_R R/I = 0\} = \{{}_RI \subseteq R \mid S\gamma(I) = S\}$.

Now let M be a left R-module and let $\alpha: M \to S \otimes_R M$ be the R-homomorphism given by $m \mapsto 1 \otimes m$. If $m \in \ker(\alpha)$ then $S \otimes_R Rm = 0$ and so $Rm \in \mathcal{J}_\tau$. Therefore $\ker(\alpha) \in \mathcal{J}_\tau$.

Let $x \in S \otimes_R M$ and let $\bar{x}$ be the image of x under the canonical map $S \otimes_R M \to \text{coker}(\alpha)$. Consider the exact sequence of left S-modules $0 \to S \otimes_R M\alpha \xrightarrow{\beta} S \otimes_R [Rx + M\alpha] \to S \otimes_R R\bar{x} = 0$. We claim that β is in fact a surjection.

Indeed, if $x = \Sigma(s_i \otimes m_i)$ then $1 \otimes x = \Sigma(1 \otimes s_i) \otimes m_i$. Since γ is a ring epimorphism, we see, as in the proof of Proposition 17.7, that $1 \otimes s_i = s_i \otimes 1$ for each i and so $x = \Sigma[(s_i \otimes 1) \otimes m_i] = \Sigma[s_i \otimes (1 \otimes m_i)] \in \mathrm{im}(\beta)$. Therefore, β is a surjection and so, by exactness, $S \otimes_R Rx = 0$, proving that $\mathrm{coker}(\alpha) \in \mathcal{T}_\tau$.

Now let N be any left S-module. Then for any $M \in \mathcal{T}_\tau$, we have $\mathrm{Hom}_R(M,N) \cong \mathrm{Hom}_R(M,\mathrm{Hom}_S(S,N)) \cong \mathrm{Hom}_R(S \otimes_R M, N) = 0$ and so ${}_RN \in \mathcal{F}_\tau$.

Moreover, if N is a left S-module, if E is the injective hull of ${}_RN$, and if E' is the injective hull of N in S-mod, then by the equivalence of (1) and (2) we have that E' is injective as a left R-module and so $E \subseteq E'$. Therefore $E/N \subseteq E'/N$ which, being a left S-module, belongs to $\mathcal{F}_\tau$ since for any $x \in E'/N$, we have that $S \otimes_R Sx = 0$ implies $Sx = 0$ and so $x = 0$. Therefore $E/N \in \mathcal{F}_\tau$.

We now apply Proposition 6.3 and see that for every left R-module M we have an isomorphism $\delta_M: S \otimes_R M \to Q_\tau(M)$ making the diagram

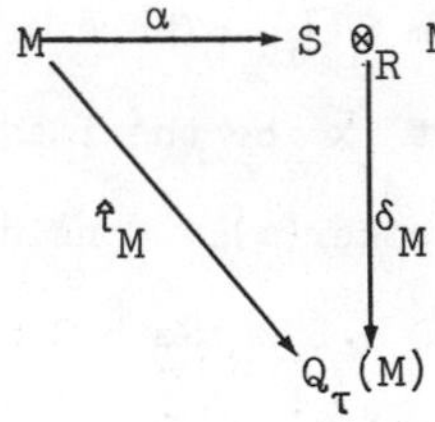

commute. This isomorphism is, moreover, unique and so we have a natural equivalence between $Q_\tau(\_)$ and $S \otimes_R \_$. That $\tau \in$ R-perf then follows from Proposition 17.1.

(3) ⇒ (1): Since τ is perfect, it is exact by Proposition 17.1 and so, by (3), $S \otimes_R \_$ is exact. This proves that S_R is flat. If $s \in S$ then, by (3), there exists an $I \in \mathcal{L}_\tau$ with $Is \subseteq \gamma(R)$. By Proposition 17.1, $\tau \in$ R-fin and so we can assume that I is finitely generated, say by $r_1,\ldots,r_n$. Then $S\gamma(I) = S$ and so $1 = \Sigma s_i\gamma(r_i)$ for $s_i \in S$. Then $S\gamma(R) = S$ and so $S \otimes_R S/\gamma(R) = 0$. Thus we have the canonical epimorphisms $S \to S \otimes_R R \to S \otimes_R \gamma(R) \to S \otimes_R S$. This suffices to show that the map $\theta\colon S \otimes_R S \to S$ defined in Proposition 17.7(5) must be an isomorphism, proving that γ is a ring epimorphism. □

(17.9) COROLLARY: <u>Let</u> $\tau \in$ R-tors. <u>Then</u> τ <u>is perfect if and only if</u> $\hat{\tau}$ <u>is a left flat ring epimorphism</u>.

PROOF: This follows directly from Proposition 17.8. □

We can apply Proposition 17.8 to show the existence of the "flat epimorphic hull" of a ring R.

(17.10) PROPOSITION: <u>For every ring</u> R <u>there exists a ring bijection</u> $\gamma\colon R \to S$ <u>satisfying the</u>

following conditions:

(1) S_R *is flat*.

(2) *If* $\gamma'\colon R \to S'$ *is a ring bijection with* S'_R *flat, then there exists a unique ring homomorphism* δ *making the diagram*

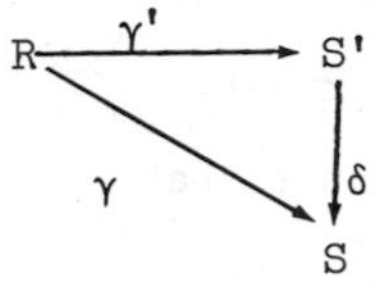

commute.

PROOF: Let U be the set of all faithful elements of R-perf. Then U is nonempty since $\xi \in U$. By Propositions 10.1 and 17.4, $\vee U \in U$. Let $S = R_{\vee U}$ and let $\gamma = \widehat{\vee U}\colon R \to S$. By Proposition 17.8, γ is a left flat ring epimorphism. Since $\vee U$ is faithful, γ is also a ring monomorphism.

Now assume that we have $\gamma'\colon R \to S'$ satisfying (2). Then by Proposition 17.8, there exists a $\tau \in$ R-perf with $S' \cong R_\tau$ and $\gamma' = \hat{\tau}$. Since γ' is a ring monomorphism, τ is faithful and so $\tau \in U$. Then $\tau \leq \vee U$ and so the δ we seek exists by Proposition 7.3. □

A great advantage of working with perfect torsion theories is that localization with respect to $\tau \in$ R-perf preserves many important properties of modules, rings, and other torsion theories. We shall now investigate some of these.

(17.11) PROPOSITION: *Let* $\tau \in$ R-perf *and let* N *be a left* R_τ*-module. Then*

(1) N *is injective as a left* R_τ*-module if and only if* N *is injective as a left* R*-module*.

(2) N *is indecomposable as a left* R_τ*-module if and only if* N *is indecomposable as a left* R*-module*.

(3) N *is uniform as a left* R_τ*-module if and only if* N *is uniform as a left* R*-module*.

PROOF: (1) This follows directly from Proposition 6.7.

(2) If N is indecomposable as a left R-module then it is clearly indecomposable as a left R_τ-module. Conversely, let N_1 and N_2 be two R-submodules of N with $N = N_1 \oplus N_2$. Then $N \cong R_\tau \otimes_R N \cong R_\tau \otimes_R N_1 \oplus R_\tau \otimes_R N_2$ and so by the indecomposability of N, we have $R_\tau \otimes_R N_i = 0$ for $i = 1$ or $i = 2$. Therefore $N_i \in \mathcal{T}_\tau$. Since N, being a left R_τ-module, is τ-torsion-free as a left R-module, this implies that $N_i = 0$.

(3) If N is uniform as a left R-module then it is clearly uniform as a left R_τ-module. Conversely, assume that N is uniform as a left R_τ-module. If E is the injective

hull of N in R_τ-mod then E is, therefore, indecomposable. By (1), E is also the injective hull of N in R-mod, and by (2), E is indecomposable in R-mod. Therefore, N is uniform as a left R-module. □

(17.12) PROPOSITION: *Let* $\tau \in$ R-perf. *If* M *is a projective left* R-*module, then* $Q_\tau(M)$ *is a projective left* R_τ-*module*.

PROOF: Let $\alpha: N \to N'$ be an R_τ-epimorphism and let $\beta: Q_\tau(M) \to N'$ be an R_τ-homomorphism. Then, considering these as maps in R-mod, there exists an R-homomorphism φ making the diagram

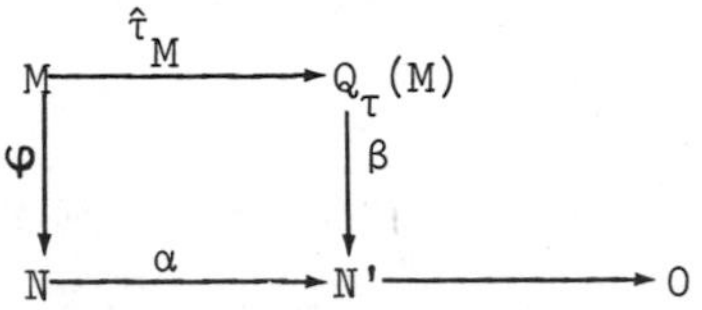

commute. Applying the exact functor $Q_\tau(\underline{\varphi})$ and noting that N and N' are absolutely τ-pure, we see that $Q_\tau(\varphi)\alpha = \beta$, proving that $Q_\tau(M)$ is projective. □

(17.13) PROPOSITION: *If* $\tau \in$ R-perf *then*

(1) *If* R *is left noetherian so is* R_τ.

(2) *If* R *is left artinian so is* R_τ.

(3) *If* R *is left hereditary so is* R_τ.

(4) *If* R_τ *is left semihereditary so is* R_τ.

(5) <u>If</u> R <u>is left semiartinian so is</u> R_τ.

PROOF: (1) and (2) follow directly from Proposition 17.5. (3) follows from Propositions 17.5 and 17.12.

(4) Let H be a finitely generated left ideal of R and let $I = \hat{\tau}^{-1}(H)$. If $\{h_1,\ldots,h_n\}$ is a generating set of H, then each $h_i = \sum_j (s_j \otimes a_{ij})$ for $s_j \in R_\tau$ and $a_{ij} \in I$. Let $I = \sum_{i,j} Ra_{ij}$. Then I' is a finitely generated left ideal of R and so, by hypothesis, is projective. Moreover, $H \subseteq R_\tau \otimes_R I' \subseteq R_\tau \otimes_R I \subseteq H$ and so $H = R_\tau \otimes_R I'$ is projective by Proposition 17.12. Therefore, R_τ is left semihereditary.

(5) Let N be a nonzero left R_τ-module. Then by hypothesis, ${}_RN$ has a simple R-submodule N'. Moreover, ${}_RN$ is τ-torsion-free by Proposition 17.1 and so N' is τ-torsion-free. Therefore, $Q_\tau(N')$ is a simple left R_τ-module by Proposition 6.4, proving that R_τ is left semiartinian. □

(17.14) PROPOSITION: <u>If</u> $\tau \in$ R-perf <u>then</u> $\hat{\tau}_\#$: gen(τ) $\to R_\tau$-tors <u>is bijective. In particular</u>, $\sigma' = \hat{\tau}_\#(\tau')$ <u>if and only if</u> $\tau' = \hat{\tau}^\#(\sigma')$.

PROOF: By Proposition 9.7, $\hat{\tau}_\#$ is monic. Since $\tau \in$ R-perf, then R_τ is flat as a right R-module and so $\hat{\tau}^\#$: R_τ-tors $\to$ R-tors is defined. In particular, if $\sigma' \in R_\tau$-tors and if $\tau' = \hat{\tau}^\#(\tau')$ then

$\mathcal{J}_{\tau'} = \{M \in R\text{-mod} \mid R_\tau \otimes_R M \in \mathcal{J}_{\sigma'}\}$. If $M \in \mathcal{J}_\tau$, then $R_\tau \otimes_R M \cong Q_\tau(M) = 0$ and so $\mathcal{J}_\tau \subseteq \mathcal{J}_{\tau'}$. Thus $\tau' \in \text{gen}(\tau)$.

Finally, $\hat{\tau}_\#(\tau') = \{N \in R_\tau\text{-mod} \mid N \in \mathcal{J}_{\tau'}\} = \{N \in R_\tau\text{-mod} \mid R_\tau \otimes_R N \in \mathcal{J}_{\sigma'}\} = \mathcal{J}_{\sigma'}$ (using the fact that $\hat{\tau}$ is a ring epimorphism). Therefore, $\hat{\tau}_\#(\tau') = \sigma'$. This shows that $\hat{\tau}_\#$ is epic and hence bijective. □

(17.15) PROPOSITION: Let $\tau \in R\text{-perf}$, $\tau' \in \text{gen}(\tau)$, and $\sigma' = \hat{\tau}_\#(\tau')$. Then

(1) $\mathcal{L}_{\tau'} = \{{}_R I \subseteq R \mid Q_\tau(I) \cong R_\tau \otimes_R I \in \mathcal{L}_{\sigma'}\}$.

(2) $\mathcal{L}_{\sigma'} = \{{}_{R_\tau} H \subseteq R_\tau \mid \hat{\tau}^{-1}(H) \in \mathcal{L}_{\tau'}\}$.

Moreover, for every left R_τ-module N,

(3) $T_{\sigma'}(N) = T_{\tau'}(N)$.

(4) N is σ'-torsion-free if and only if ${}_R N$ is τ'-torsion-free.

(5) N is σ'-injective if and only if ${}_R N$ is τ'-injective.

(6) N is absolutely σ'-pure if and only if ${}_R N$ is absolutely τ'-pure.

PROOF: (1) This follows from the fact that $\tau' = \hat{\tau}^\#(\sigma')$.

(2) This follows from Proposition 9.6.

(3) Directly from the definition, we have $T_{\sigma'}(N) \subseteq T_{\tau'}(N)$. Conversely, if $x \in T_{\tau'}(N)$ then $I = \{r \in R \mid rx = 0\} \in \mathcal{L}_{\tau'}$, and so if $H = Q_\tau(I)$ then

$H \in \mathcal{L}_{\sigma'}$ by (1). Since $Hx = 0$, we have $\{s \in R_\tau \mid sx = 0\} \in \mathcal{L}_{\sigma'}$ and so $x \in T_{\sigma'}(N)$, establishing equality.

(4) This is a direct consequence of (3).

(5) Assume that N is σ'-injective. Let $I \in \mathcal{L}_{\tau'}$ and let $\alpha: I \to N$ be an R-homomorphism. Define the R_τ-homomorphism $\alpha': R_\tau \otimes_R I \to N$ by $\alpha': \Sigma(s_i \otimes a_i) \mapsto \Sigma s_i(a_i\alpha)$. Then by (1), $R_\tau \otimes_R I \in \mathcal{L}_{\sigma'}$ and so, by the σ'-injectivity of N, α' can be extended to an R_τ-homomorphism $\beta': R_\tau \to N$. Let $x = 1\beta'$. Then $r \mapsto rx$ is an R-homomorphism $R \to N$ that extends α and so proves that ${}_RN$ is τ'-injective.

Conversely, let ${}_RN$ be τ'-injective. Let $H \in \mathcal{L}_{\sigma'}$ and let $\alpha: H \to N$ be an R_τ-homomorphism. Set $I = \hat{\tau}^{-1}(H)$. By (2), $I \in \mathcal{L}_{\tau'}$. Define an R-homomorphism $\alpha': I \to N$ by $a \mapsto \hat{\tau}(a)\alpha$. Since ${}_RN$ is τ-injective, α' extends to an R-homomorphism $\beta': R \to N$. Let $x = 1\beta'$. Then if $\Sigma(s_i \otimes a_i) \in R_\tau \otimes_R I = H$, we have $[\Sigma(s_i \otimes a_i)]\alpha = \Sigma s_i(a_i\alpha') = \Sigma s_i a_i x$ and so the R_τ-homomorphism $R_\tau \to N$ defined by $s \mapsto sx$ extends α.

(6) This follows from (4) and (5). □

(17.16) PROPOSITION: <u>Let</u> $\tau \in$ R-perf, $\tau' \in$ gen(τ), <u>and</u> $\sigma' = \hat{\tau}_\#(\tau')$. <u>Then a left</u> R-<u>module</u> M <u>is</u>

τ'-*projective if and only if* $Q_\tau(M)$ *is* σ'-*projective*.

PROOF: Assume that M is τ'-projective. If $\alpha\colon N \to N'$ is an R_τ-epimorphism with $N \in \mathcal{E}_{\sigma'}$ and $N' \in \mathcal{F}_{\sigma'}$, then by Proposition 17.15 we have ${}_RN \in \mathcal{E}_{\tau'}$ and ${}_RN' \in \mathcal{F}_{\tau'}$. The proof that $Q_\tau(M)$ is σ'-projective then follows in the same manner as the proof of Proposition 17.12.

Conversely, assume that $Q_\tau(M)$ is σ'-projective. Let $\alpha\colon N \to N'$ be an R-epimorphism with $N, N' \in \mathcal{F}_{\tau'}$ and let $\beta\colon M \to N'$ be an R-homomorphism. Since $N' \in \mathcal{F}_{\tau'} \subseteq \mathcal{F}_\tau$, we have $T_\tau(M)\beta = 0$ and so $0 \neq \beta$ induces an R-homomorphism $M/T_\tau(M) \to N'$. Therefore, to prove that M is τ'-projective, it will clearly suffice to prove that $M/T_\tau(M)$ is τ'-projective. Therefore, without loss of generality, assume that $M \in \mathcal{F}_\tau$.

Since τ is perfect, it is exact and so $Q_\tau(\alpha)\colon Q_\tau(N) \to Q_\tau(N')$ is an R_τ-epimorphism with $Q_\tau(N)$, $Q_\tau(N') \in \mathcal{F}_{\sigma'}$. By Proposition 16.2, there then exists a σ'-dense submodule M' of $Q_\tau(M)$ and an R_τ-homomorphism making the following diagram commute.

$$\begin{array}{ccccc}
0 & \longrightarrow & M' & \longrightarrow & Q_\tau(M) \\
 & & \Big\downarrow \varphi & & \Big\downarrow Q_\tau(\beta) \\
 & & Q_\tau(N) & \xrightarrow{\;Q_\tau(\alpha)\;} & Q_\tau(N')
\end{array}$$

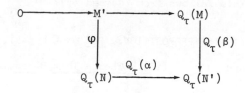

Note that M' cannot be equal to 0 since, if it were, then $Q_\tau(M)$ would be σ'-torsion whence $Q_\tau(\beta)$ would be the zero map. Moreover, since $M \in \mathcal{F}_\tau$, we have that $\hat{\tau}_M$ is a monomorphism. Therefore, without loss of generality, we can consider M as a large submodule of $Q_\tau(M)$. Thus $0 \neq M_0 = M' \cap M$ and $M/M_0 \cong [M + M']/M' \subseteq Q_\tau(M)/M'$, and so M_0 is τ'-dense in M.

If $\varphi = 0$ then $M_0 Q_\tau(\beta) = 0$, and so $M_0\beta = 0$. Therefore, the diagram

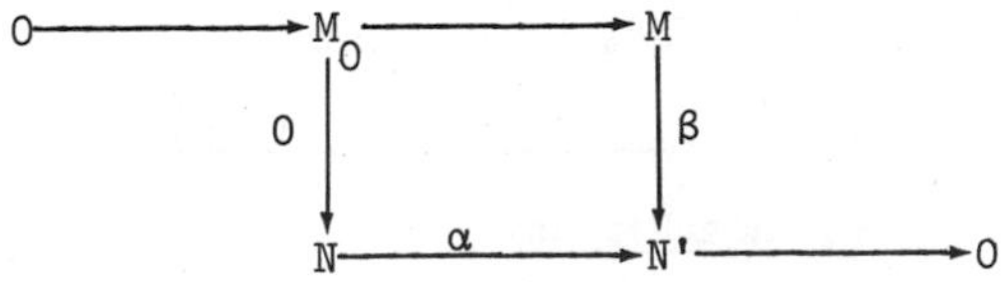

commutes and we are done. Assume, therefore, that $\varphi \neq 0$. Since $N \in \mathcal{F}_{\tau'}$, we have in particular that $N \in \mathcal{F}_\tau$ and so we can assume, without loss of generality, that N is a large submodule of $Q_\tau(N)$. Thus $0 \neq N_1 = M'\varphi \cap N$, and so $0 \neq M_1 = N_1\varphi^{-1} \cap M_0$. Moreover, the diagram

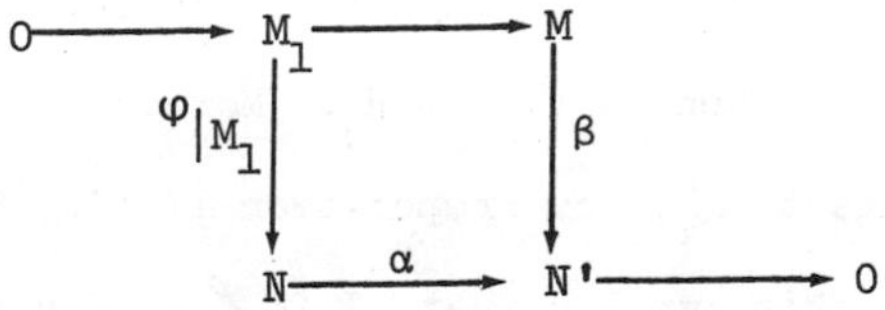

commutes. We are therefore done if we can show that M_1 is τ'-dense in M.

From the exact sequence $0 \to M_0/M_1 \to M/M_1 \to M/M_0 \to 0$ and the fact that M_0 is τ'-dense in M, we see that it suffices to show that M_1 is τ'-dense in M_0. Consider the R-homomorphism $M_0 \xrightarrow{\psi} Q_\tau(N) \xrightarrow{\nu} Q_\tau(N)/N$ where $\psi = \varphi|_{M_0}$ and ν is the canonical projection. If $x\psi\nu = 0$ then $x\psi \in N$ and so $x \in N\psi^{-1} = M_1$. Therefore, $\psi\nu$ induces an R-monomorphism $M_0/M_1 \to Q_\tau(N)/N \in \mathcal{T}_\tau \subseteq \mathcal{T}_{\tau'}$, proving that M_1 is τ'-dense in M_0. □

(17.18) PROPOSITION: Let $\tau \in$ R-perf, $\tau' \in \text{gen}(\tau)$, and $\sigma' = \hat{\tau}_\#(\tau')$. Then

(1) τ' is noetherian if and only if σ' is noetherian.

(2) τ' is of finite type if and only if σ' is of finite type.

(3) τ' is exact if and only if σ' is exact.

(4) τ' is perfect if and only if σ' is perfect.

PROOF: (1) Let τ' be noetherian and let $H_1 \subseteq H_2 \subseteq \ldots$ be an ascending chain of left ideals of R_τ with $H = \cup H_j \in \mathcal{L}_{\sigma'}$. For each positive integer j, let $I_j = \hat{\tau}^{-1}(H_j)$. Then $Q_\tau(I_j) = H_j$. Moreover, $\cup I_j = \hat{\tau}^{-1}(H)$, which belongs to $\mathcal{L}_{\tau'}$ by Proposition 17.15. Since τ' is noetherian, this implies that $I_k \in \mathcal{L}_{\sigma'}$ for some positive integer k and so $H_k \in \mathcal{L}_{\sigma'}$, proving that σ' is noetherian. The converse is proven similarly.

(2) Let τ' be of finite type. If $H \in \mathcal{L}_{\sigma'}$, set $I = \hat{\tau}^{-1}(H)$. Then $I \in \mathcal{L}_{\tau'}$ and so there exists a finitely generated $I' \in \mathcal{L}_{\tau'}$ containing I. Let $H' = Q_\tau(I') \cong R_\tau \otimes_R I'$. Then H' is finitely generated and belongs to $\mathcal{L}_{\sigma'}$, proving that σ' is of finite type.

Conversely, assume that σ' is of finite type and let $I \in \mathcal{L}_{\tau'}$. Then $Q_\tau(I) \in \mathcal{L}_{\sigma'}$, and so there exists a finitely generated $H' \in \mathcal{L}_{\sigma'}$ contained in $Q_\tau(I)$. Let $\{h_1,\ldots,h_n\}$ be a generating set of H', where each $h_i = \Sigma(s_{ij} \otimes a_{ij})$ for some $s_{ij} \in R_\tau$ and $a_{ij} \in I$. If $I' = \sum_{i,j} Ra_{ij}$, then I' is a finitely generated left ideal of R contained in I and $Q_\tau(I') = R_\tau \otimes_R I' \supseteq H'$. Therefore, $Q_\tau(I') \in \mathcal{L}_{\sigma'}$ and so $I' \in \mathcal{L}_{\tau'}$. This proves that τ' is of finite type.

(3) Assume that τ' is exact. If $H \in \mathcal{L}_{\sigma'}$ and $I = \hat{\tau}^{-1}(H)$, then $I \in \mathcal{L}_{\tau'}$ and so I is τ'-projective. By Proposition 17.16, $H \cong Q_\tau(I)$ is σ'-projective. Thus σ' is exact by Proposition 16.3. The converse is proven similarly.

(4) This follows from (1), (3), and Proposition 17.1(8). □

In Section 13 we studied the characterization of coatoms of R-tors. For perfect torsion theories we can sharpen the result we obtained in Proposition 13.7.

(17.19) PROPOSITION: *For* $\tau \in$ R-perf *the following conditions are equivalent:*

(1) τ *is a coatom of* R-tors.

(2) (i) R_τ *is left local;*

(ii) $J(R_\tau)$ *is right* T-*nilpotent; and*

(iii) $R_\tau/J(R_\tau)$ *is simple artinian.*

(3) (i) τ *is saturated; and*

(ii) *every nonzero injective left* R_τ*-module is faithful.*

PROOF: (1) ⇒ (2): By Proposition 17.14, R_τ-tors = $\{\xi,\chi\}$ and so (2i) and (2ii) follow from Propositions 12.2 and 12.9. By Proposition 12.9, $R_\tau/J(R_\tau)$ is left semiartinian, left local, and left primitive, and so (2iii) follows from Proposition 12.14.

(2) ⇒ (3): If (2) is satisfied, then any nonzero injective left R_τ-module is τ-torsion-free as a left R-module and is faithful over R_τ. This implies (3) by Proposition 13.7.

(3) ⇒ (1): This follows from Proposition 13.7. □

Finally, we will find it useful later on to be able to reduce the study of perfect torsion theories to that of faithful perfect torsion theories. To be able to do this we need the following result.

(17.20) PROPOSITION: Let $\tau \in R\text{-tors}$ and let I be a two-sided ideal of R contained in $T_\tau(R)$. Let $\gamma: R \to R/I = S$ be the canonical ring surjection and let $\sigma = \gamma_\#(\tau)$. Then τ is perfect if and only if σ is perfect.

PROOF: Let N be a left S-module. Since $I \subseteq T_\tau(R)$, we have $IQ_\tau(N) \subseteq T_\tau(R)Q_\tau(N) \subseteq T_\tau(Q_\tau(N)) = 0$ and so $Q_\tau({}_RN)$ is canonically a left S-module. Now assume that τ is perfect. By Proposition 9.9, $Q_\tau({}_RN) \cong Q_\sigma({}_SN)$ and so, in particular, $S_\sigma \cong Q_\sigma(S) \cong Q_\tau({}_RS) \cong R_\tau \otimes_R S \cong R_\tau/R_\tau I$. Thus, if W is a left S_σ-module, then W is a left R_τ-module via the induced ring homomorphism $R_\tau \to R_\tau/R_\tau I \cong S_\sigma$ and so $T_\tau(W) = 0$. Therefore, $T_\sigma(W) = 0$ and so σ is perfect by Proposition 17.1(5).

Conversely, suppose σ is perfect. Then $S_\sigma \cong Q_\sigma({}_SS) \cong Q_\tau({}_RS) \cong Q_\tau(R/T_\tau(R)) \cong R_\tau$. This induces an isomorphism between R_τ-mod and S_σ-mod. In particular, every left R_τ-module belongs to $\mathcal{F}_\tau$ and so τ is perfect by Proposition 17.1(5). □

In particular, if $\tau \in R\text{-perf}$, if $\gamma: R \to R/T_\tau(R) = S$ is the canonical surjection, and if $\sigma = \gamma_\#(\tau)$, then by Proposition 17.20, $\sigma \in S\text{-perf}$ and σ is faithful.

References for Section 17

Beachy [10]; Cheatham [29]; Goldman [63]; Hacque [66, 68]; Lambek [90, 92]; Morita [105, 107, 108, 109]; Nastasescu and Popescu [113]; Page [117]; Popescu and Spircu [123]; Raynaud [125]; Rubin [129]; Silver [142]; Stenström [146, 147]; Storrer [148, 150]; Walker and Walker [164]; Winton [165].

CHAPTER IV

THE LEFT SPECTRUM

18. COCRITICAL MODULES

Let $\tau \in R\text{-tors}$. A nonzero left R-module M is said to be τ-*cocritical* if and only if M is τ-torsion-free and every nonzero submodule of M is τ-dense in M. For example, a simple left R-module is τ-cocritical for every τ relative to which it is torsion-free.

(18.1) PROPOSITION: *The following conditions are equivalent for a nonzero left R-module M:*

(1) M *is* $\chi(M)$-*cocritical.*

(2) *There exists a* $\tau \in R\text{-prop}$ *with* M τ-*cocritical.*

PROOF: (1) ⇒ (2): The proof is trivial.

(2) ⇒ (1): If M is a τ-cocritical left R-module for some $\tau \in R\text{-prop}$, then $M \in \mathcal{F}_\tau$ and so $\tau \leq \chi(M)$. Thus $\mathcal{J}_\tau \subseteq \mathcal{J}_{\chi(M)}$ and so every nonzero submodule of M, being τ-dense in M, is also $\chi(M)$-dense in M. Therefore, M is $\chi(M)$-cocritical. □

We say that a nonzero left R-module M is <u>cocritical</u> if and only if it is $\chi(M)$-cocritical.

(18.2) PROPOSITION: <u>Let</u> $\tau \in$ R-tors <u>and let</u> M <u>be a</u> τ-<u>cocritical left</u> R-<u>module. Then</u>

(1) <u>Every nonzero submodule of</u> M <u>is</u> τ-<u>cocritical</u>.

(2) <u>If</u> $\tau' \leq \tau$ <u>then</u> $Q_{\tau'}(M)$ <u>is</u> τ-<u>cocritical</u>.

(3) <u>If</u> $\tau = \chi(N)$ <u>then</u> N <u>has a nonzero</u> τ-<u>cocritical submodule</u>.

(4) M <u>is uniform</u>.

(5) <u>If</u> $N \in \mathcal{F}_\tau$ <u>then every nonzero</u> R-<u>homomorphism</u> $M \to N$ <u>is a monomorphism</u>.

(6) <u>The endomorphism ring of</u> M <u>is embeddable in a division ring</u>.

PROOF: (1) Let N be a nonzero submodule of M. Then N is also τ-torsion-free. If N' is a nonzero submodule of N then N/N' is a submodule of M/N' and so is τ-torsion. Thus N is τ-cocritical.

(2) Since $\tau' \leq \tau$, $M \in \mathcal{F}_\tau \subseteq \mathcal{F}_{\tau'}$. Therefore, we can consider M as a large submodule of $Q_{\tau'}(M)$ which implies that $Q_{\tau'}(M) \in \mathcal{F}_\tau$. Let $0 \neq {}_RN \subseteq Q_{\tau'}(M)$. Then by the largeness of M, we have $0 \neq N \cap M$. Moreover, we have the exact sequence $0 \to M/[M \cap N] \to Q_{\tau'}(M)/N \to Q_{\tau'}(M)/[M+N] \to 0$. Since M is τ-cocritical, $M/[M \cap N]$ is τ-torsion. Since $\tau' \leq \tau$, $Q_{\tau'}(M)/M \subseteq Q_\tau(M)/M$ and so $Q_{\tau'}(M)/M$ is τ-torsion.

This implies that $Q_{\tau'}(M)/[M + N]$ is τ-torsion and so we conclude that $Q_{\tau'}(M)/N$ is τ-torsion. Therefore, $Q_{\tau'}(M)$ is τ-cocritical.

(3) Since M is τ-torsion-free there exists a nonzero R-homomorphism $\alpha: M \to E(N)$. Since M is τ-cocritical and $E(N)$ is τ-torsion-free, α must be a monomorphism. Therefore, by (1), $0 \neq M\alpha \cap N$ is a τ-cocritical submodule of N.

(4) Suppose we have nonzero submodules N and N' of M with $0 = N \cap N'$. Then N is τ-torsion-free since it is a submodule of M. On the other hand, N is isomorphic to a submodule of M/N' and so is τ-torsion, which is a contradiction.

(5) If $N \in \mathcal{F}_\tau$ and $\alpha: M \to N$ is a nonzero R-homomorphism, then $M/\ker(\alpha)$ is isomorphic to a submodule of N and so is τ-torsion-free. Since M is τ-cocritical, this implies that $\ker(\alpha) = 0$.

(6) Let S be the endomorphism ring of M. By (2), $E_\tau(M)$ is also τ-cocritical. If N is a submodule of $E_\tau(M)$ and if $\alpha \in \mathrm{Hom}_R(N,E_\tau(M))$, then N is τ-dense in $E_\tau(M)$ and so by τ-injectivity there exists an endomorphism β of $E_\tau(M)$ extending α. Thus $E_\tau(M)$ is quasi-injective and so contains the quasi-injective hull M' of M (see [44]). In particular, M' is also τ-cocritical.

Let S' be the endomoprhism ring of M'. If $0 \neq \alpha \in S'$, then by (5) α is a monomorphism. Moreover, we have an R-homomorphism $\beta: \mathrm{im}(\alpha) \to M'$ such that $\alpha\beta$ is the identity on M'. Since M' is quasi-injective, β extends to an endomorphism β' of M'. If $x \in M'$ then $x\beta' = m\alpha\beta$ for some $m \in M'$. But by (5), β' is an R-monomorphism and so $x = m\alpha$. Therefore, α is an R-epimorphism and so an isomorphism. This proves that S' is a division ring.

By Proposition 5.1, every endomorphism α of M can be uniquely extended to an endomorphism α' of $E_\tau(M)$. By the properties of the quasi-injective hull, $M'\alpha' \subseteq M'$ and so this extension, followed by restriction to M', gives us a function $S \to S'$ that is easily seen to be a ring homomorphism. Moreover, if β is an endomorphism of M' with $M\beta = 0$, then β induces an R-homomorphism $M'/M \to M'$. Since M' is τ-cocritical, M'/M is τ-torsion and M' is τ-torsion-free. Therefore, we must have $\beta = 0$. This suffices to show that S is embedded in S'. □

Note that the converse of Proposition 18.2(4) is false, i.e., there exist uniform left R-modules that are not cocritical. For an example, see Chapter VII, Example 29.

(18.3) COROLLARY: Let $\tau \in$ R-tors. Then the following conditions are equivalent for a nonzero left R-module M:

(1) M is τ-cocritical.

(2) Rm <u>is</u> τ-<u>cocritical for every</u> $0 \neq m \in M$.

PROOF: (1) $\Rightarrow$ (2): This is proven by Proposition 18.2(1).

(2) $\Rightarrow$ (1): By (2), Rm is τ-torsion-free for all $m \in M$ and so M is τ-torsion-free. Assume that M is not τ-cocritical. Then there exists a nonzero proper τ-pure submodule N of M. Pick $m \in M \setminus N$. Then we have a canonical monomorphism $Rm/[Rm \cap N] \to M/N$ and so $Rm \cap N$ is a proper submodule of Rm that is nonzero by Proposition 18.2(4). Furthermore, $Rm \cap N$ is then τ-pure in Rm, and so Rm is not τ-cocritical. □

The next result shows that cocriticalness, like the many other properties studied in Section 17, is preserved by localization at a perfect torsion theory.

(18.4) PROPOSITION: <u>Let</u> $\tau \in R\text{-perf}$, $\tau' \in \text{gen}(\tau)$, <u>and</u> $\sigma' = \hat{\tau}_{\#}(\tau')$. <u>Then</u>

(1) $M \in \mathcal{F}_{\tau'}$ <u>is</u> τ'-<u>cocritical if and only if</u> $Q_{\tau}(M)$ <u>is</u> σ'-<u>cocritical</u>.

(2) $N \in \mathcal{F}_{\sigma'}$ <u>is</u> σ'-<u>critical if and only if</u> ${}_RN$ <u>is</u> τ'-<u>cocritical</u>.

PROOF: (1) Let M be a τ'-cocritical left R-module and let N be a proper σ'-pure submodule of $Q_{\tau}(M)$ with $M' = N\hat{\tau}_M^{-1}$. Then $N = Q_{\tau}(M)$ and $Q_{\tau}(M)/N \cong Q_{\tau}(M/M')$.

By Proposition 17.15(4), $M/M' \in \mathcal{F}_{\tau'}$ and so by τ'-cocriticalness $M' = 0$ and hence $N = 0$. Thus $Q_\tau(M)$ is σ'-cocritical.

Conversely, let $Q_\tau(M)$ be σ'-cocritical and let M' be a proper τ'-pure submodule of M. Then $Q_\tau(M)/Q_\tau(M') \cong Q_\tau(M/M') \in \mathcal{F}_{\sigma'}$, by the definition of σ' and so $Q_\tau(M') = 0$. Since M' is a submodule of M, it is τ'-torsion-free and hence τ-torsion-free and so this implies that $M' = 0$.

(2) This is proven similarly. □

(18.5) COROLLARY: If $\tau \in$ R-perf, then a left R_τ-module N is simple if and only if N is τ-cocritical.

PROOF: This follows from Proposition 18.4, using the fact that $\hat{\tau}_\#(\tau) = \xi$ and that a module is ξ-cocritical if and only if it is simple. □

Corollary 18.3 suggests that we can restrict our attention to cyclic cocritical left R-modules. Indeed, by Proposition 18.2(4), $E(M) = E(Rm)$ for any cocritical left R-module M and any $0 \neq m \in M$. We say that a left ideal I of R is τ-critical if and only if the cyclic left R-module R/I is τ-cocritical for $\tau \in$ R-tors. Note that I is τ-critical if and only if it is a maximal element of $\mathcal{C}_\tau$. We say that the left ideal I of R is <u>critical</u> if and only if it is $\chi(R/I)$-critical.

(18.6) PROPOSITION: *If I is a two-sided ideal of a ring R that is critical as a left ideal of R, then I is completely prime.*

PROOF: Let $a,b \in R$ with $ab \in I$ and $a \notin I$. Set $H = Ra + I$. Since $I \subset H$, we have R/H is $\chi(R/I)$-torsion and so there exists no nonzero R-homomorphism $R/H \to R/I$. In particular, the map defined by $r + H \mapsto rb + I$ must be the zero map, which implies that $b \in I$. □

(18.7) PROPOSITION: *The following conditions are equivalent for a two-sided ideal I of a ring R:*

(1) I *is a critical left ideal of* R.

(2) R/I *is a left Öre domain.*

PROOF: (1) ⇒ (2): By Proposition 18.6, (1) implies that I is completely prime and so R/I is an integral domain. Moreover, if $a,b \in R$ with $a \in R \setminus I$, then $I \subset (I + Ra:b)$ and so there exist $a' \in (I + Ra:b) \setminus I$ and $b' \in R$ with $a'b - b'a \in I$. Thus R/I is a left Öre domain.

(2) ⇒ (1): To prove (1), it suffices to show that for every left ideal H of R properly containing I, we have $H \in \mathcal{L}_{\chi(R/I)}$. Indeed, it suffices to consider H of the form $Ra + I$ where $a \in R \setminus I$. Therefore, pick $a \in R \setminus I$ and

let $H = Ra + I$. Assume that we have a nonzero R-homomorphism $R/H \to E(R/I)$. Then there exists $b \in R \setminus H$ with $(b + H)\alpha = c + I \in R/I$. Therefore, $(H:b) \subseteq (I:c)$ and $(I:c) = I$ since R/I is an integral domain. But by the left Öre condition there exists an $a' \in R \setminus I$ and a $b' \in R$ with $a'b - b'a \in I$ and so $(H:b) \not\subseteq I$. From this contradiction we see that there can be no nonzero R-homomorphisms $R/H \to E(R/I)$ and so $H \in \mathcal{L}_{\chi(R/I)}$. □

Note that as a consequence of Proposition 18.7, we see that R is a left Öre domain if and only if 0 is a critical ideal of R.

(18.8) COROLLARY: <u>If</u> R <u>is commutative then an ideal</u> I <u>of</u> R <u>is critical if and only if it is prime</u>.

PROOF: This follows directly from Proposition 18.7 since, for a commutative ring R, an ideal I of R is prime if and only if R/I is an integral domain. □

Recall the well-known result that a left R-module M is uniform if and only if $E(M)$ is indecomposable.

(18.9) PROPOSITION: <u>Let</u> $\tau \in$ R-tors. <u>If</u> I <u>is a</u> τ-<u>critical left ideal of</u> R <u>then so is</u> $(I:a)$ <u>for any</u> $a \in R \setminus I$. <u>Moreover</u>, $E(R/I) \cong E(R/(I:a))$.

PROOF: By Proposition 18.2(4), R/I is uniform and so $E(R/I)$ is indecomposable. Define the R-homomorphism $\alpha: R/(I:a) \to R/I$ by $\alpha: r + (I:a) \mapsto ra + I$. Then α is a monomorphism since $r \in (I:a)$ if and only if $ra \in I$. Therefore $(I:a)$ is τ-critical by Proposition 18.2(1). Moreover, α can then be extended to an R-homomorphism $\beta: E(R/(I:a)) \to E(R/I)$ which must also be a monomorphism since $R/(I:a)$ is large in its injective hull. Since $E(R/I)$ is indecomposable, $\mathrm{im}(\beta)$ must equal $E(R/I)$ and so β is in fact an isomorphism. □

(18.10) PROPOSITION: The following conditions are equivalent for a left ideal I of R:

(1) I is critical.

(2) R/I is uniform and I is a maximal element of $\{{}_RH \subseteq R \mid R/H$ is uniform and $E(R/H) \cong E(R/I)\}$.

(3) There exists an indecomposable injective left R-module M for which I is a maximal element of $\{(0:m) \mid 0 \neq m \in M\}$.

PROOF: (1) ⇒ (2): That R/I is uniform follows by Proposition 18.2. Now assume that H is a left ideal of R properly containing I. Since I is critical, $R/H \in \mathcal{T}_{\chi(R/I)}$ and so we cannot possibly have $E(R/H) \cong E(R/I)$. Thus we have (2).

(2) ⇒ (3): By (2), $M = E(R/I)$ is an indecomposable injective left R-module and $I = (0:1+I)$. Suppose that $I \subseteq (0:m)$ for some $0 \neq m \in M$. Then by indecomposability, $M = E(Rm) \cong E(R/(0:m))$ and so, by (2), $(0:m) = I$.

(3) ⇒ (1): By the indecomposability of M, we have $M = E(R/I)$. If H is a left ideal of R properly containing I and if $\alpha \in \mathrm{Hom}_R(R/H,M)$, then let $x = (1 + H)\alpha$. Then $(0:x) \supseteq H \supset I$ and so by (2) we must have $(0:x) = R$. Therefore, $x = 0$ and so $\alpha = 0$. This shows that $R/H \in \mathcal{J}_{\chi(R/I)}$ and so that I is critical. □

For any $\tau \in$ R-tors, set $K(\tau) = \cap\{{}_R I \subseteq R \mid I$ is τ-critical$\}$ (if there are no τ-critical left ideals of R set $K(\tau) = R$). Since each τ-critical left ideal I of R belongs to $\mathcal{C}_\tau$ and $\mathcal{C}_\tau$ is closed under intersections, we have $K(\tau) \in \mathcal{C}_\tau$.

(18.11) PROPOSITION: <u>For</u> $\tau \in$ R-tors, $K(\tau) = \cap\{(0:M) \mid M \in$ R-mod <u>is</u> τ-<u>cocritical</u>$\}$ = $\cap\{(0:M) \mid M \in$ R-mod <u>is</u> τ-<u>cocritical and</u> τ-<u>injective</u>$\}$.

PROOF: If M is a τ-cocritical left R-module, then $(0:m)$ is a τ-critical left ideal of R for every $0 \neq m \in M$ by Corollary 18.3, and so $(0:M) = \cap\{(0:m) \mid 0 \neq m \in M\} \supseteq K(\tau)$. Therefore, $\cap\{(0:M) \mid M \in$ R-mod is τ-cocritical$\} \supseteq K(\tau)$.

Conversely, if a belongs to this intersection, then, in particular, for every τ-critical left ideal I of R, we have $a \in (0:R/I) \subseteq I$ and so $a \in K(\tau)$. This establishes the first equality.

If M is a τ-cocritical left R-module, then by Proposition 18.2(2) $Q_\tau(M)$ is also τ-cocritical. Since M is τ-torsion-free, M is a submodule of $Q_\tau(M)$ and so $(0:M) \supseteq (0:Q_\tau(M))$. Therefore, $\cap\{(0:M) \mid M \in R\text{-mod}$ is τ-cocritical and τ-injective$\} \subseteq K(\tau)$. The reverse containment is trivial and so we have the second equality. □

In particular, we note that for any $\tau \in R\text{-tors}$, $K(\tau)$ is a two-sided ideal of R.

(18.12) PROPOSITION: _If_ $\tau \in R\text{-perf}$, _then_

(1) $K(\tau) = \hat{\tau}^{-1}(J(R_\tau))$.

(2) $J(R_\tau) = R_\tau[\hat{\tau}(K(\tau))]$.

PROOF: (1) Let M be a simple left R_τ-module. By Corollary 18.5, M is τ-cocritical as a left R-module and so, by Proposition 18.11, $K(\tau) \subseteq (0:M) = \hat{\tau}^{-1}(H)$ where $H = \{s \in R_\tau \mid sM = 0\}$. Since $J(R_\tau)$ is the intersection of all annihilators of simple left R_τ-modules, we have $K(\tau) \subseteq \hat{\tau}^{-1}(J(R_\tau))$. Conversely, for each τ-critical left ideal I of R, $Q_\tau(R/I)$ is a simple left R_τ-module. Thus for any $r \in \hat{\tau}^{-1}(J(R_\tau))$, we have $\hat{\tau}(r)Q_\tau(R/I) = 0$ and so

$r \in (0:Q_\tau(R/I)) \subseteq (0:R/I)$. Since this is true for each such I, $r \in K(\tau)$ and so we have equality.

(2) By (1), $J(R_\tau) \supseteq R_\tau[\hat{\tau}(K(\tau))]$. Conversely, let $s \in J(R_\tau)$ and let $H = \{r \in R \mid rs \in \hat{\tau}(R)\}$. Then $\hat{\tau}(H)s \subseteq \hat{\tau}(R) \cap J(R_\tau) = \hat{\tau}(K(\tau))$. Since $H \in \mathcal{L}_\tau$ and τ is perfect, $R_\tau\hat{\tau}(H) = R_\tau$ and so $s \in R_\tau s = R_\tau\hat{\tau}(H)s \subseteq R_\tau[\hat{\tau}(K(\tau))]$. Thus we have the reverse inclusion and hence equality. □

It is not, in general, true that every $\tau \in R$-prop has a τ-cocritical left R-module. See Chapter VII, Example 28. If this condition does hold, we say that R is *left seminoetherian*.

(18.13) PROPOSITION: *If* R *is left semiartinian then it is left seminoetherian.*

PROOF: Let R be a left semiartinian and let $\tau \in R$-prop. If M is a nonzero τ-torsion-free left R-module, then M has a nonzero simple submodule M' which must clearly be τ-cocritical. □

We can then add the following to Proposition 17.13.

(18.14) PROPOSITION: *If* $\tau \in R$-perf *and if* R *is left seminoetherian, then* R_τ *is left seminoetherian.*

PROOF: Let $\sigma' \in R_\tau$-prop and let $\tau' = \hat{\tau}^{\#}(\sigma')$.

Then there exists a τ'-cocritical left R-module M and so $Q_\tau(M)$ is σ'-cocritical by Proposition 18.4. □

We now define a chain $\{\tau_i\}$ of torsion theories on R-mod as follows:

(i) $\tau_{-1} = \xi$.

(ii) if i is not a limit ordinal, $\tau_i = \tau_{i-1} \vee \xi(\{M \in \text{R-mod} \mid M \text{ is } \tau_{i-1}\text{-cocritical}\})$.

(iii) if i is a limit ordinal, $\tau_i = \vee\{\tau_j \mid j < i\}$.

This chain is called the <u>Gabriel filtration</u> on R-mod. If $M \in$ R-mod, we say that M has <u>Gabriel dimension</u> k if and only if the set of ordinals i for which $M \in \mathcal{T}_{\tau_i}$ is nonempty and k is its minimal member. The ring R is said to have <u>left Gabriel dimension</u> k if and only if ${}_RR$ has Gabriel dimension k. It is immediate that R has left Gabriel dimension k if and only if $\tau_k = \chi$ and $\tau_i \neq \chi$ for any $i < k$.

(18.15) PROPOSITION: <u>The following conditions are equivalent for a ring</u> R:

(1) R <u>is left seminoetherian</u>.

(2) R <u>has left Gabriel dimension</u> k <u>for some ordinal</u> k.

PROOF: Let $\{\tau_i\}$ be the Gabriel filtration on R-mod.

(1) ⇒ (2): Assume that R is left seminoetherian. Since R-tors is a set, there exists an ordinal k that is minimal with respect to $\tau_k = \tau_{k+1}$. If $\tau_k \neq \chi$, there exists a τ_k-cocritical module M. Then by definition $M \in \mathcal{T}_{\tau_k} \setminus \mathcal{T}_{\tau_{k+1}}$, which is a contradiction. Thus we must have $\tau_k = \chi$ and so R has left Gabriel dimension.

(2) ⇒ (1): Let $\tau \in$ R-prop and let i be the least ordinal such that $\tau_i \not\leq \tau$. (Such an ordinal exists since $\tau_k = \chi \not\leq \tau$). Then i is not a limit ordinal. By the definition of τ_i and the minimality of i, there exists a τ_{i-1}-cocritical left R-module M satisfying $M \in \mathcal{T}_{\tau_i} \setminus \mathcal{T}_{\tau}$. Hence for any nonzero submodule N of M, $M/N \in \mathcal{T}_{\tau_{i-1}} \subseteq \mathcal{T}_{\tau}$. Since M is not τ-torsion, this implies that M must be τ-torsion-free and so M is τ-cocritical. Thus every such τ has a τ-cocritical left R-module and so R is left seminoetherian. □

(18.16) PROPOSITION: *Let* R *be a left seminoetherian ring and let* $\tau < \tau'$ *be proper torsion theories on* R-mod. *Then there exists a* τ-*cocritical left* R-*module that is* τ'-*torsion*.

PROOF: Since R is left seminoetherian, the class of τ-cocritical left R-modules is nonempty. Assume that no τ-cocritical left R-module is τ'-torsion and let $\{\tau_i\}$ be the Gabriel filtration on R-mod. Let h be the least ordinal satisfying $\tau_h \not\leq \tau$. Then h is not a limit ordinal.

We claim that $\tau_i \wedge \tau' \leq \tau$ for every ordinal i. If $i < h$ then $\tau_i \leq \tau$ and so the result is immediate. We therefore consider an ordinal i and assume that for all $j < i$ we have $\tau_j \wedge \tau' \leq \tau$. Here we distinguish two cases.

Case I: i is a limit ordinal. Let $M \in \mathcal{J}_{\tau_i} \cap \mathcal{J}_{\tau'}$ and assume that M is not τ-torsion. Replacing M by $M/T_\tau(M)$ if necessary, if suffices to consider M τ-torsion-free. For any $j < i$ let $N \in \mathcal{J}_{\tau_j}$. If $0 \neq \alpha \in \mathrm{Hom}_R(N,E(M))$ then $N\alpha \cap M \in \mathcal{J}_{\tau_j} \cap \mathcal{J}_{\tau'} \subseteq \mathcal{J}_{\tau'}$, which is a contradiction. Thus $\mathrm{Hom}_R(N,E(M)) = 0$ and so $M \in \mathcal{F}_{\tau_j}$ for all $j < i$. This implies that $M \in \mathcal{F}_{\tau_i}$, which is a contradiction. Therefore $\tau_i \wedge \tau' \leq \tau$.

Case II: i is not a limit ordinal. Let $M \in \mathcal{J}_{\tau_i} \cap \mathcal{J}_{\tau'}$ and assume that M is not τ-torsion. Again we can assume without loss of generality that M is τ-torsion-free. Moreover, by induction, M is not τ_{i-1}-torsion and so, replacing M by $M/T_{\tau_{i-1}}(M)$ if necessary, we can assume that M is τ_{i-1}-torsion-free. Since $M \in \mathcal{J}_{\tau_i} \cap \mathcal{F}_{\tau_{i-1}}$, there exists a τ_{i-1}-cocritical left R-module N and a nonzero R-homomorphism $\alpha: N \to M$. Since M is τ_{i-1}-torsion-free, α must in fact be a monomorphism by Proposition 18.2 and so we can assume that $N \subseteq M$. We claim that N is in fact τ-cocritical. Indeed, if N' is a nonzero submodule of N

then $N/N' \in \mathcal{J}_{\tau_{i-1}} \cap \mathcal{J}_{\tau'} \subseteq \mathcal{J}_{\tau}$. On the other hand, $N \subseteq M \in \mathcal{F}_{\tau}$. But then N is a τ-cocritical τ'-torsion left R-module, which is a contradiction. Therefore $\tau_i \wedge \tau' \leq \tau$.

Thus the claim is established. But R is left seminoetherian and so, by Proposition 18.15, there exists an ordinal k for which $\tau_k = \chi$ whence $\tau' = \tau_k \wedge \tau' \leq \tau$, which is a contradiction. This proves the theorem. □

References for Section 18

Gabriel [52]; Goldie [62]; Goldman [63]; Hudry [74, 75, 76]; Lambek and Michler [94]; Nastasescu and Popescu [112, 113]; Popescu [119, 120, 121, 122]; Raynaud [126, 127]; Sim [143, 144]; Storrer [149].

19. PRIME TORSION THEORIES

Let $\tau \in R$-tors. If $\tau = \chi(M)$ for some cocritical left R-module M, then we say that τ is prime. This choice of terminology is motivated by Proposition 18.7 and the fact that by the uniformness of M, we have $E(M) \cong E(Rm)$ for every $0 \neq m \in M$ and so $\chi(M) = \chi(Rm)$ for any $0 \neq m \in M$. Thus τ is prime if and only if it is the torsion theory cogenerated by R/I where I is a critical left ideal of R. If M is a simple left R-module, then $\chi(M)$ is clearly prime.

The set of all prime members of R-tors will be called the left spectrum of R and will be denoted by R-sp.

(19.1) PROPOSITION: A ring R is left local if and only if ξ is prime.

PROOF: If R is left local, then R-simp = {M} for some left R-module M and so, by Proposition 12.3, $\xi = \chi(M)$. Since M is simple, ξ is then prime.

Conversely, assume that ξ is prime. Then $\xi = \chi(M)$ for some cocritical left R-module M, which clearly must be simple. If M' is a simple left R-module not isomorphic to M, then by Proposition 12.2, $\xi(M') \leq \chi(M) = \xi$, which is a contradiction. Therefore, R is left local. □

To begin the study of prime torsion theories we must consider the properties of left ideals of R that are critical with respect to them.

(19.2) PROPOSITION: Let $\tau \in$ R-sp. Then the following conditions are equivalent for a cocritical left R-module M:

(1) M is τ-cocritical.

(2) $\tau = \chi(M)$.

PROOF: (1) ⇒ (2): Since τ is prime, $\tau = \chi(M')$ for some cocritical left R-module M'. Since M' is uniform, $E(M') \in \tau$ is indecomposable. Since M is τ-cocritical, by Proposition 18.2 there exists an R-monomorphism $M \to E(M')$ that can be extended by injectivity to an R-monomorphism

$E(M) \to E(M')$. By the indecomposability of $E(M')$, this is in fact an isomorphism, and so $E(M) \in \tau$, whence $\tau = \chi(M)$.

(2) ⇒ (1). This follows from Proposition 18.1. □

(19.3) PROPOSITION: *Let* $\tau \in R\text{-sp}$. *Then the following conditions are equivalent for a left ideal* I *of* R:

(1) $I \notin \mathcal{L}_\tau$.

(2) *There exists an* $a \in R \setminus I$ *such that* $(I:a)$ *is contained in a* τ-*critical left ideal of* R.

PROOF: (1) ⇒ (2): Since τ is prime, $\tau = \chi(R/H)$ for some τ-critical left ideal H of R. If $I \notin \mathcal{L}_\tau$ then there exists a nonzero R-homomorphism $\alpha: R/I \to E(R/H)$. Let $x = (1 + I)\alpha$. Then there exists an $a \in R$ with $0 \neq ax \in R/H$. By Proposition 18.2, $(0:ax)$ is a τ-critical left ideal of R and $(I:a) \subseteq (0:ax)$.

(2) ⇒ (1): If $I \in \mathcal{L}_\tau$, then $(I:a) \in \mathcal{L}_\tau$ for all $a \in R$. Therefore, any left ideal of R containing $(I:a)$ belongs to $\mathcal{L}_\tau$ and so cannot possibly be τ-critical. □

(19.4) PROPOSITION: *For* $\tau, \tau' \in R\text{-sp}$ *the following conditions are equivalent*:

(1) $\tau' \leq \tau$.

(2) *If* I *and* I' *are a* τ-*critical and a* τ'-*critical left ideal of* R, *respectively*,

<u>then there exists an</u> $a \in R \setminus I$ <u>and an</u> $a' \in R \setminus I'$ <u>with</u> $(I:a) \subseteq (I':a')$.

PROOF: (1) $\Rightarrow$ (2): Since $\tau' \leq \tau$, we see that R/I is τ'-torsion-free and so there exists a nonzero R-homomorphism $\alpha: R/I \to E(R/I')$. Since R/I' is large in its injective hull, $im(\alpha) \cap R/I' \neq 0$ and so there exists an $a \in R \setminus I$ with $0 \neq (a + I)\alpha = a' + I' \in R/I'$. This implies that $(I:a) \subseteq (I':a')$.

(2) $\Rightarrow$ (1): Since $\tau = \chi(R/I)$ by Proposition 19.2, to show that $\tau' \leq \tau$ it suffices to show that R/I is τ'-torsion-free. To do this it suffices to show that $(I:a) \notin \mathcal{L}_{\tau'}$ for any $a \in R \setminus I$.

Let $a \in R \setminus I$. By Proposition 18.9, $(I:a)$ is also τ-critical and so by (2) there exists a $b \in R \setminus (I:a)$ and an $a' \in R \setminus I'$ such that $((I:a):b) \subseteq (I':a')$. Since $(I':a')$ is τ'-critical by Proposition 18.9, it follows from Proposition 19.3 that $(I:a) \notin \mathcal{L}_{\tau'}$. □

(19.5) PROPOSITION: <u>If</u> $\tau \in$ R-sp <u>and if</u> I, I' <u>are</u> τ-<u>critical left ideals of</u> R, <u>then there exist</u> $a \in R \setminus I$ <u>and</u> $a' \in R \setminus I'$ <u>with</u> $(I:a) = (I':a')$.

PROOF: By Proposition 19.2, $\chi(R/I) = \tau = \chi(R/I')$. Therefore there exists a nonzero R-homomorphism $\alpha: R/I \to E(R/I')$. By Proposition 18.2, this is in fact a monomorphism. As in the first part of the proof of

Proposition 19.4, there exist $a \in R \smallsetminus I$ and $a' \in R \smallsetminus I'$ with $(I:a) \subseteq (I':a')$ where, in fact, we have equality since α is monic. □

If $\tau \in R\text{-sp}$ then there exists a τ-cocritical left R-module M. By Proposition 18.2(2), $Q_\tau(M)$ is also τ-cocritical and, moreover, is absolutely τ-pure. The next result shows that this characterizes $Q_\tau(M)$ up to isomorphism.

(19.6) PROPOSITION: _If_ $\tau \in R\text{-sp}$ _then any two absolutely_ τ-_pure_ τ-_cocritical left_ R-_modules are isomorphic_.

PROOF: Let N and N' be absolutely τ-pure τ-cocritical left R-modules. Since N and E(N') are τ-torsion-free and since $\tau = \chi(N')$ by Proposition 19.2, there exists an R-homomorphism $\alpha: N \to E(N')$ that is a monomorphism by Proposition 18.2. Since N' is large in E(N'), $N'' = N\alpha \cap N'$ is nonzero. Since $N\alpha \cong N$, it is also τ-cocritical and so $N\alpha/N'' \in \mathcal{J}_\tau$. By τ-injectivity we, therefore, have an R-homomorphism $\beta: N\alpha \to N'$ extending the embedding $N'' \to N'$. Since N'' is large in $N\alpha$, β is also a monomorphism. Since N' is τ-cocritical, $N'/N\alpha\beta \in \mathcal{J}_\tau$. On the other hand, $N\alpha\beta$ is absolutely τ-pure since N is, and N' is τ-torsion-free so we must have $N'/N\alpha\beta \in \mathcal{F}_\tau$. Therefore, $N' = N\alpha\beta$ and so $N' \cong N$. □

We note that if $\tau \in R\text{-sp}$ and if M is an absolutely τ-pure τ-cocritical left R-module, then by Propositions 18.11 and 19.6, $(0:M) = K(\tau)$.

We now turn to a characterization of perfect prime torsion theories.

(19.7) PROPOSITION: <u>The following conditions are equivalent for</u> $\tau \in R\text{-tors}$:

(1) τ <u>is perfect and prime</u>.

(2) R_τ <u>is left local and if</u> $\tau = \chi(M)$ <u>then</u> $Q_\tau(M)$ <u>has a nonzero socle in</u> R_τ-mod.

(3) τ <u>is perfect and</u> R_τ <u>is left local</u>.

PROOF: (1) ⇒ (2): Let N and N' be simple left R_τ-modules. Since τ is perfect, N and N' are absolutely τ-pure as left R-modules by Proposition 17.1. By Corollary 18.5, N and N' are τ-cocritical as left R-modules and so, by Proposition 19.6, $N \cong N'$ as R-modules and hence as R_τ-modules. Therefore R_τ is left local.

If $\tau = \chi(M)$, then by Proposition 18.3, M has a τ-cocritical submodule M'. Then $Q_\tau(M')$ is also τ-cocritical by the same proposition, and so by Corollary 18.5, $Q_\tau(M')$ is a simple left R_τ-module. Therefore, $Q_\tau(M)$ has a nonzero socle in R_τ-mod.

(2) ⇒ (3): Since R_τ is left local, all simple left R_τ-modules are isomorphic. Thus if N is a simple left

R_τ-module and $\tau = \chi(M)$, then by (2), N is isomorphic to a submodule of $Q_\tau(M)$ and so, as a left R-module, is τ-torsion-free. Then (3) follows by Proposition 17.1(7).

(3) ⇒ (1): Let M be a simple left R_τ-module. Since τ is perfect, M is absolutely τ-pure as a left R-module. Since all simple left R_τ-modules are isomorphic, E(M) is an injective cogenerator of R_τ-mod and so every left R_τ-module, considered as a left R-module, is $\chi(M)$-torsion-free. This implies that every τ-torsion-free left R-module is $\chi(M)$-torsion-free and so $\chi(M) \leq \tau$. But M is τ-torsion-free as a left R-module, so $\tau \leq \chi(M)$, proving equality. By Corollary 18.5, M is τ-cocritical as a left R-module and so, by definition, τ is prime. □

(19.8) PROPOSITION: <u>The following conditions are equivalent for</u> $\tau \in R\text{-sp} \cap R\text{-perf}$:

(1) R_τ <u>is a simple ring</u>.

(2) $T_\tau(R) = K(\tau)$.

PROOF: By Proposition 19.7, R_τ is left local. As we saw in Section 12, $J(R_\tau)$ is therefore the unique maximal two-sided ideal of R_τ, and so R_τ is simple if and only if $J(R_\tau) = 0$. By Proposition 18.12, this happens if and only if $K(\tau) \subseteq \ker(\hat{\tau}) = T_\tau(R)$. But $T_\tau(R)$ is always contained in $K(\tau)$, by the definition of $K(\tau)$, and so we have the equivalence of (1) and (2). □

(19.9) PROPOSITION: <u>The following conditions are equivalent for</u> $\tau \in$ R-tors:

(1) $S = R_\tau / J(R_\tau)$ <u>is a simple left artinian ring</u> and ${}_R S \in \mathcal{F}_\tau$.

(2) (i) $\tau \in$ R-sp $\cap$ R-perf; <u>and</u>

(ii) $\tau = \chi(R/K(\tau))$.

PROOF: (1) $\Rightarrow$ (2): Since $J(R_\tau)$ annihilates all simple left R_τ-modules, every simple left R_τ-module is also a simple left S-module. Therefore, all simple left R_τ-modules are isomorphic, proving that R_τ is left local. Moreover, each simple left R_τ-module is isomorphic to an R_τ-submodule M of S, and so is τ-torsion-free as a left R-module. By Proposition 17.1(7), this proves that $\tau \in$ R-perf and so we have (2i).

By Proposition 18.12, $\hat{\ }$ induces an R-monomorphism $R/K(\tau) \to S$, and so $\chi(R/K(\tau)) \geq \chi(S)$. On the other hand, $Q_\tau(R/K(\tau)) \cong Q_\tau(R)/Q_\tau(K(\tau)) \cong R_\tau/J(R_\tau) = S$ by Proposition 18.12, so, indeed, $\chi(R/K(\tau)) = \chi(S)$. Moreover, by Corollary 18.5, M is τ-cocritical as a left R-module and so we have $\tau = \chi(M) \geq \chi(S) \geq \tau$ which proves (2ii).

(2) $\Rightarrow$ (1): Since $\tau = \chi(R/K(\tau))$ and $S \cong Q_\tau(R/K(\tau))$, then by Proposition 19.7, S has a nonzero socle in R_τ-mod and so has a nonzero socle in S-mod. Thus S, being a simple ring with a nonzero socle, is simple and left artinian.

Moreover, since τ is perfect and S is a left R_τ-module, we have ${}_RS \in \mathcal{F}_\tau$. □

We now turn to consider the structure of the set R-sp. Note that the partial order on R-tors induces a partial order on R-sp. However, R-sp need not be a sublattice of R-tors, as we shall see later.

(19.10) PROPOSITION: *If* M *is a simple left* R-*module, then* $\chi(M)$ *is a minimal element of* R-sp.

PROOF: We have already noted that for a simple left R-module M, $\chi(M)$ is prime. Suppose $\xi < \tau \leq \chi(M)$. Then $M \in \mathcal{F}_\tau$ and so M is τ-critical. By Proposition 19.2, we then have $\tau = \chi(M)$ if τ is prime. Therefore, $\chi(M)$ is a minimal element of R-sp. □

(19.11) PROPOSITION: *Let* $\tau \in$ R-sp *and let* $\tau_1, \tau_2 \in$ R-tors. *Then*

(1) $\tau = \tau_1 \wedge \tau_2$ *implies that* $\tau = \tau_1$ *or* $\tau = \tau_2$.

(2) $\tau \geq \tau_1 \wedge \tau_2$ *implies that* $\tau \geq \tau_1$ *or* $\tau \geq \tau_2$.

PROOF: (1) Assume that $\tau = \tau_1 \wedge \tau_2$, where $\tau_1, \tau_2 > \tau$. If I is a τ-critical left ideal of R, then $\tau = \chi(R/I)$ by Proposition 19.2, and so $I \notin \mathcal{C}_{\tau_1} \cup \mathcal{C}_{\tau_2}$. For $j = 1,2$ let H_j be the τ_j-purification of I in R. Then $0 \neq H_j/I$ for $j = 1,2$. On the other hand, $(H_1/I) \cap (H_2/I) = T_{\tau_1 \wedge \tau_2}(R/I) = T_\tau(R/I) = 0$ and so $I = H_1 \cap H_2$. This contradicts Proposition 18.10(2).

Assume that $\tau \geq \tau_1 \wedge \tau_2$ and define $\tau_j' = \tau_j \vee \tau$ for $j = 1,2$. Then clearly $\tau \leq \tau_1' \wedge \tau_2'$. Assume that this inequality is strict. Then there exists a nonzero left R-module M belonging to $(\mathcal{J}_{\tau_1'} \cap \mathcal{J}_{\tau_2'}) \setminus \mathcal{J}_\tau$. Replacing M by $0 \neq M/T_\tau(M)$ if necessary, we can in fact assume that $M \in \mathcal{F}_\tau$. Since $M \in \mathcal{J}_{\tau_1'}$ we must have that $M \notin \mathcal{F}_{\tau_1}$. Therefore, replacing M by $0 \neq T_{\tau_1}(M)$ if necessary, we can assume that $M \in \mathcal{J}_{\tau_1} \cap \mathcal{F}_\tau$. Similarly, we can assume that $M \in \mathcal{J}_{\tau_2} \cap \mathcal{F}_\tau$. But then $0 \neq M \in \mathcal{J}_{\tau_1} \cap \mathcal{J}_{\tau_2} \cap \mathcal{F}_\tau \subseteq \mathcal{J}_\tau \cap \mathcal{F}_\tau$, which is a contradiction. Thus $\tau = \tau_1' \wedge \tau_2'$. Applying (1), we see that $\tau = \tau_j'$ for $j = 1$ or $j = 2$, which proves that $\tau \geq \tau_j$ for that j. □

As we remarked earlier, if $U \subseteq$ R-sp then $\wedge U$ need not belong to R-sp. Indeed, we are far from that situation, as the next result shows.

(19.12) PROPOSITION: _Let_ $U \subseteq$ R-sp. _Then_ $\wedge U \in$ R-sp _if and only if_ U _has a unique minimal member_.

PROOF: If U has a unique minimal member τ, then $\wedge U = \tau \in$ R-sp. Conversely, assume that $\wedge U \in$ R-sp. To show that U has a unique minimal member, it suffices to show that $\wedge U \in U$. Assume that this is not the case. Then

$\wedge U < \tau$ for every $\tau \in U$. Since $\wedge U \in R\text{-sp}$, $\wedge U = \chi(R/I)$ for some critical left ideal I of R. If $\tau \in U$ and $R/I \in \mathcal{F}_\tau$, then $\tau \leq \chi(R/I) = \wedge U$, which is a contradiction. Therefore, I is not τ-pure in R for any $\tau \in U$. For each $\tau \in U$, let H_τ be the τ-purification of I in R. We claim that in fact $H_\tau = R$ for all such τ, for else we would have $R/H_\tau \in \mathcal{F}_\tau \subset \mathcal{F}_{\wedge U}$, contradicting the fact that I is $\wedge U$-critical. Therefore, R/I is τ-torsion for all $\tau \in U$ and so $0 = T_{\wedge U}(R/I) = \cap\{T_\tau(R/I) \mid \tau \in U\} = R/I$, which is a contradiction. Thus we must have $\wedge U \in U$. □

If $\tau' \leq \tau$ are prime torsion theories, then by Proposition 19.4 there exist a τ-critical left ideal I of R and a τ'-critical left ideal I' of R with $I \subseteq I'$. If the converse holds, we say that R is <u>left balanced</u>. That is to say, R is left balanced if and only if, for critical left ideals $I \subseteq I'$ of R, we have $\chi(R/I) \geq \chi(R/I')$. If R is a left duo ring (i.e., a ring for which every left ideal is two-sided) then R is clearly left balanced.

(19.13) PROPOSITION: <u>A sufficient condition for</u> R <u>to be left balanced is that every</u> $\tau \in R\text{-sp}$ <u>be stable</u>.

PROOF: Assume that every $\tau \in R\text{-sp}$ is stable. If $I \subseteq I'$ are critical left ideals of R, then $R/I \notin \mathcal{T}_{\chi(R/I')}$

and so $E(R/I) \notin \mathcal{J}_{\chi(R/I')}$. By the cocriticalness of R/I, $E(R/I)$ is indecomposable, and so by Proposition 11.3, $E(R/I) \in \mathcal{F}_{\chi(R/I')}$. Therefore, $\chi(R/I) \geq \chi(R/I')$. □

(19.14) PROPOSITION: *If* R *is a left balanced ring, then the following conditions are equivalent for* $\tau \in R\text{-sp}$.

(1) $\tau = \chi(M)$ *for some simple left* R-*module* M.

(2) τ *is a minimal element of* R-sp.

(3) *Every* τ-*critical left ideal of* R *is a maximal left ideal of* R.

PROOF: (1) ⇒ (2): This follows from Proposition 19.10.

(2) ⇒ (3): Let I be a τ-critical left ideal of R. Then I is contained in a maximal left ideal H of R. Since $\chi(R/H)$ is prime and R is left balanced, $\tau = \chi(R/I) \geq \chi(R/H)$. By (2), this implies that $\tau = \chi(R/I) = \chi(R/H)$. In particular, R/H is τ-torsion-free and so we must have $I = H$.

(3) ⇒ (1): The proof is trivial. □

(19.15) PROPOSITION: *The following conditions are equivalent for a ring* R:

(1) R *is left semiartinian*.

(2) (i) *Every* $\tau \in R\text{-sp}$ *has a simple* τ-*cocritical module*;

(ii) $\chi(E)$ _is prime for every indecomposable injective left_ R-_module_ E; _and_

(iii) _Every nonzero left_ R-_module contains a nonzero uniform submodule_.

PROOF: (1) ⇒ (2): If $\tau \in$ R-sp and if M is a τ-cocritical left R-module, then by (1), M has a nonzero simple submodule M' that is also τ-cocritical by Proposition 18.2. This proves (2i). If E is an indecomposable injective left R-module and M is a simple submodule of E, then $E = E(M)$ and so $E \in \chi(M) \in$ R-sp, proving (2ii). Finally, (2iii) follows directly from Proposition 18.2.

(2) ⇒ (1): Let M be a left R-module. By (2), M contains a nonzero uniform submodule N. Since E(N) is an indecomposable injective left R-module, $\chi(N)$ is prime, so there exists a simple $\chi(N)$-cocritical left R-module M'. In particular, there exists a nonzero R-homomorphism $\alpha: M' \to E(N)$ which must be a monomorphism by the simplicity of M'. Since N is large in E(N), we have $M'\alpha \subseteq N$ and so $M'\alpha$ is a simple submodule of M. Thus R is left semiartinian. □

(19.16) PROPOSITION: _If_ R _is left semiartinian, then every prime torsion theory is minimal in_ R-sp.

PROOF: Let I be a critical left ideal of R and let $\tau = \chi(R/I)$. Then $E(R/I)$ is an indecomposable injective left R-module. Since R is left semiartinian, R/I has a simple submodule N, and by the indecomposability of $E(R/I)$, we have $E(N) = E(R/I)$ and so $\tau = \chi(N)$. The result then follows by Proposition 19.10. □

(19.17) PROPOSITION: <u>If</u> R <u>is left noetherian, then</u> R-sp <u>satisfies the descending-chain condition</u>.

PROOF: Let $\tau_1 \geq \tau_2 \geq \ldots$ be a descending chain of members of R-sp. Let I_1 be a τ_1-critical left ideal of R. By Proposition 19.2, $\tau_1 = \chi(R/I_1)$ and so $I_1 \notin \mathcal{L}_{\tau_1}$. Thus $I_1 \notin \mathcal{L}_{\tau_2}$. Since R is left noetherian, $\tau_2 \in$ R-fin. Since $I_1 \notin \mathcal{L}_{\tau_2}$, the τ_2-purification of I_1 in R is a proper left ideal of R, and so by Proposition 15.2 it (and hence I_1) is contained in a τ_2-critical left ideal I_2. Continue in this manner to obtain an ascending chain $I_1 \subseteq I_2 \subseteq \ldots$ of left ideals of R with $\tau_i = \chi(R/I_i)$ for each positive integer i. Since R is left noetherian, there is then a positive integer k with $I_k = I_{k+1} = \ldots$ and so $\tau_k = \tau_{k+1} = \ldots$ □

To each left R-module M, we associate the subset $\text{supp}(M) = \{\tau \in \text{R-sp} \mid M \notin \mathcal{T}_\tau\}$ of R-sp, which we call the support of M. Then $\text{supp}(M) = \{\tau \in R\text{-sp} \mid Q_\tau(M) \neq 0\}$. Moreover, the support satisfies the properties that we expect.

(19.18) PROPOSITION: For a left R-module M:

(1) If $M = \Sigma M_i$, then $\text{supp}(M) = \bigcup \text{supp}(M_i)$.

(2) If N is a submodule of M, then

$\text{supp}(M) = \text{supp}(N) \cup \text{supp}(M/N)$.

PROOF: The proof follows directly from the definition. □

In particular, we note that for any $U \subseteq R\text{-sp}$, we have $\mathcal{J}_{\wedge(R\text{-sp} \setminus U)} = \{M \in R\text{-mod} \mid \text{supp}(M) \subseteq U\}$.

Similarly, with each $\tau \in R\text{-prop}$, we associate the subset $\text{pgen}(\tau) = \text{gen}(\tau) \cap R\text{-sp}$ of R-sp consisting of all prime generalizations of τ.

(19.19) PROPOSITION: If $\tau, \tau' \in R\text{-prop}$, then $\text{pgen}(\tau \wedge \tau') = \text{pgen}(\tau) \cup \text{pgen}(\tau')$.

PROOF: That $\text{pgen}(\tau) \cup \text{pgen}(\tau') \subseteq \text{pgen}(\tau \wedge \tau')$ follows from Proposition 8.3. The reverse inclusion follows from Proposition 19.11. □

(19.20) PROPOSITION: If R is left balanced and $U \subseteq R\text{-sp}$, then $\text{pgen}(\wedge U) = \bigcup\{\text{pgen}(\tau) \mid \tau \in U\}$.

PROOF: That $\bigcup\{\text{pgen}(\tau) \mid \tau \in U\} \subseteq \text{pgen}(\wedge U)$ follows from Proposition 8.3. Conversely, let $\tau \in \text{pgen}(\wedge U)$. If I is a τ-critical left ideal of R, then $R/I \in \mathcal{F}_\tau \subseteq \mathcal{F}_{\wedge U}$ and so there exists a $\tau' \in U$ with $I \notin \mathcal{L}_{\tau'}$. By Proposition 19.3, this means that there exist an $a \in R \setminus I$ and a τ'-critical left ideal I' of R with $(I:a) \subseteq I'$. But $(I:a)$ is also τ-critical by Proposition 18.9. Since R is left balanced, we therefore have $\tau = \chi(R/(I:a)) \geq \chi(R/I') = \tau'$ and so $\cup\{\text{pgen}(\tau) \mid \tau \in U\} \supseteq \text{pgen}(\wedge U)$. □

The following result extends Proposition 19.10.

(19.21) PROPOSITION: <u>If</u> $\tau \in$ R-prop <u>and if</u> M <u>is a</u> τ-<u>cocritical left</u> R-<u>module, then</u> $\chi(M)$ <u>is a minimal element of</u> pgen(τ).

PROOF: Let $\tau'' \in \text{pgen}(\tau)$ and assume that $\tau'' \leq \chi(M)$. Then $M \in \mathcal{F}_{\chi(M)} \subseteq \mathcal{F}_{\tau''}$. On the other hand, for any nonzero submodule M' of M, we have $M/M' \in \mathcal{T}_\tau \subseteq \mathcal{T}_{\tau''}$. Therefore, M is τ''-cocritical and so, by Proposition 19.2, $\tau'' = \chi(M)$. □

In general, it is not true that every minimal element of pgen(τ) is of the form $\chi(M)$ where M is a τ-cocritical left R-module. When this holds we say that τ is <u>clear</u>.

(19.22) PROPOSITION: <u>If</u> R <u>is left balanced, then a sufficient condition for</u> $\tau \in$ R-prop <u>to be clear</u>

is that every $I \notin \mathcal{L}_\tau$ be contained in a τ-critical left ideal.

PROOF: Let $\tau' = \chi(R/I')$ be a minimal element of pgen(τ), where I' is a τ'-critical left ideal of R. Since $\tau \leq \tau'$, $I' \in \mathcal{C}_\tau$ and so, by hypothesis, I' is contained in a τ-critical left ideal H of R. Since R is left balanced, $\chi(R/I') \geq \chi(R/H)$ and so, by the minimality of τ', we have $\tau' = \chi(R/I') = \chi(R/H)$. Thus τ is clear. □

(19.23) PROPOSITION: If every $\tau \in$ R-prop is clear, then R is left balanced.

PROOF: Let $I \subset I'$ be critical left ideals of R and let $\tau = \chi(R/I)$ and $\tau' = \chi(R/I')$. If $\tau \not\geq \tau'$ then $\tau'' = \tau \wedge \tau'$ is different from τ'. By Proposition 19.19, $\{\tau,\tau'\}$ is precisely the set of minimal elements of pgen(τ''). Since τ'' is clear, we then have, in particular, that $\tau = \chi(N)$ for some τ''-cocritical left R-module N. Since $I \in \mathcal{C}_\tau$, we then have a nonzero R-homomorphism $R/I \to N$ that is a monomorphism by Proposition 18.2. Therefore, R/I is isomorphic to a submodule of N, and so I is τ''-critical. Since $I' \supset I$, this implies that $R/I' \in \mathcal{J}_{\tau''} \subseteq \mathcal{J}_{\tau'}$, which is a contradiction. □

Finally, we see that localization at a perfect torsion

theory preserves primeness, thus adding another condition to the list in Proposition 17.18.

(19.24) PROPOSITION: <u>Let</u> $\tau \in$ R-perf, $\tau' \in \text{gen}(\tau)$, <u>and</u> $\sigma' = \hat{\tau}_{\#}(\tau')$. <u>Then</u> τ' <u>is prime if and only if</u> σ' <u>is prime</u>.

PROOF: Assume that τ' is prime and let M be a cocritical left R-module with $\tau' = \chi(M)$. Then M is τ'-cocritical by Proposition 19.2. By Proposition 18.2, $Q_\tau(M)$ is also τ'-cocritical and so $\tau' = \chi(Q_\tau(M))$. A left R_τ-module N is, therefore, σ'-torsion if and only if $\text{Hom}(N,E(Q_\tau(M)) = 0$ (by Proposition 6.6 it is immaterial whether we consider these as R-homomorphisms or R_τ-homomorphisms) and so $\sigma' = \chi(Q_\tau(M))$. By Proposition 18.4, $Q_\tau(M)$ is σ'-critical and so σ' is prime.

Conversely, assume that σ' is prime and let N be a σ'-cocritical left R_τ-module. Then ${}_RN$ is τ'-cocritical by Proposition 18.4 and $\tau' = \chi({}_RN)$. This implies that τ' is prime. □

(19.25) COROLLARY: <u>If</u> $\tau \in$ R-perf <u>then</u> $\hat{\tau}_{\#}\colon \text{pgen}(\tau) \to R_\tau\text{-sp}$ <u>is bijective</u>.

PROOF: This follows from Propositions 17.14 and 19.21 . □

References for Section 19

Cahen [24]; Gabriel [52]; Golan [56]; Goldman [63]; Hudry [74, 75, 76]; Lambek and Michler [94]; Popescu [119, 120, 121, 122]; Raynaud [126, 127]; Sim [143, 144]; Storrer [149].

20. SEMIPRIME TORSION THEORIES

Let $\tau \in$ R-prop. We define the root of τ by $\sqrt{\tau} = \wedge \text{pgen}(\tau)$. Then $\sqrt{\tau} \geq \tau$ and in general we do not have equality.

(20.1) PROPOSITION: For $\tau, \tau' \in$ R-tors,

(1) $\tau \leq \tau'$ implies that $\sqrt{\tau} \leq \sqrt{\tau'}$

(2) $\sqrt{\sqrt{\tau}} = \sqrt{\tau}$.

(3) $\sqrt{\tau \wedge \tau'} = \sqrt{\tau} \wedge \sqrt{\tau'}$.

(4) $\sqrt{\tau \vee \tau'} = \sqrt{\sqrt{\tau} \vee \sqrt{\tau'}}$.

PROOF: The proofs of (1) and (2) are trivial. As for (3), by Proposition 19.19, we have $\text{pgen}(\tau \wedge \tau') = \text{pgen}(\tau) \cup \text{pgen}(\tau')$. Therefore, by Proposition 8.4, $\sqrt{\tau \wedge \tau'} = \wedge\text{pgen}(\tau \wedge \tau') = [\wedge\text{pgen}(\tau)] \wedge [\wedge\text{pgen}(\tau')] = \sqrt{\tau} \wedge \sqrt{\tau'}$. As for (4), $\tau \vee \tau' \leq \sqrt{\tau} \vee \sqrt{\tau'}$ implies that $\sqrt{\tau \vee \tau'} \leq \sqrt{\sqrt{\tau} \vee \sqrt{\tau'}}$ by (1). On the other hand, if $\tau'' \in \text{pgen}(\tau \vee \tau')$, then $\tau'' \in \text{pgen}(\tau) \cap \text{pgen}(\tau')$ and so $\tau'' \geq \sqrt{\tau} \vee \sqrt{\tau'}$. This implies the reverse inequality and so we have equality. □

Let $\tau \in$ R-prop. If $\tau = \sqrt{\tau}$ we say that τ is semiprime. By Proposition 12.10, if R is left semiartinian then every $\tau \in$ R-prop is semiprime.

(20.2) PROPOSITION: *If* $\tau, \tau' \in$ R-prop *are semiprime then* $\tau \wedge \tau'$ *is semiprime*.

PROOF: By Proposition 20.1,
$\sqrt{\tau \wedge \tau'} = \sqrt{\tau} \wedge \sqrt{\tau'} = \tau \wedge \tau'$. □

Note that if every $\tau \in$ R-prop is semiprime, then the function $\tau \mapsto \text{pgen}(\tau)$ is monic. Indeed, if $\text{pgen}(\tau) = \text{pgen}(\tau')$, then $\tau = \sqrt{\tau} = \wedge\text{pgen}(\tau) = \wedge\text{pgen}(\tau') = \sqrt{\tau'} = \tau'$.

(20.3) PROPOSITION: *The following conditions are equivalent for* $\tau \in$ R-prop:

(1) τ *is semiprime*.

(2) *If* $I \notin \mathcal{L}_\tau$ *there exists a* $\tau' \in \text{pgen}(\tau)$ *with* $I \notin \mathcal{L}_{\tau'}$.

(3) *If* $I \notin \mathcal{L}_\tau$ *there exists a critical left ideal* H *of* R *such that*

(i) $(I:a) \subseteq H$ *for some* $a \in R \setminus I$; *and*

(ii) $(H:b) \notin \mathcal{L}_\tau$ *for all* $b \in R \setminus H$.

PROOF: (1) ⇔ (2): This is proven trivially.

(2) ⇒ (3): Let $I \notin \mathcal{L}_\tau$ and let $\tau' \in \text{pgen}(\tau)$ satisfy $I \notin \mathcal{L}_{\tau'}$. By Proposition 19.3, there exists a τ'-critical

left ideal H of R and an $a \in R \setminus I$ with $(I:a) \subseteq H$, thus proving (3i). For any $b \in R \setminus H$, $(H:b)$ is τ'-critical by Proposition 18.9. Thus $(H:b) \in \mathcal{C}_{\tau'} \subseteq \mathcal{C}_\tau$ and so $(H:b) \notin \mathcal{L}_\tau$, proving (3ii).

(3) ⇒ (2): Let $I \notin \mathcal{L}_\tau$ and let H be a left ideal of R satisfying (3i) and (3ii). Set $\tau' = \chi(R/H)$. By (3i) and Proposition 19.3, $I \notin \mathcal{L}_{\tau'}$. To show that $\tau' \in \text{pgen}(\tau)$, it suffices to show that $H \in \mathcal{C}_\tau$. Assume that this is not so. Then $R/H \notin \mathcal{F}_\tau$ and so there exists a $b \in R \setminus H$ with $(H:b) \in \mathcal{L}_\tau$, contradicting (3ii). □

(20.4) PROPOSITION: <u>If every</u> $\tau \in$ R-prop <u>is semiprime, then the following conditions are equivalent for</u> $\tau \in$ R-prop:

(1) τ <u>is a maximal element of</u> R-prop.

(2) τ <u>is a maximal element of</u> R-sp.

PROOF: (1) ⇒ (2): If τ is a maximal element of R-prop, then $\text{gen}(\tau) = \{\tau, \chi\}$ and so $\text{pgen}(\tau) \subseteq \{\tau\}$. Since τ is semiprime, $\tau = \wedge\text{pgen}(\tau)$ and so $\text{pgen}(\tau)$ is nonempty. Therefore, τ is prime and (2) follows immediately.

(2) ⇒ (1): If $\tau' \in$ R-prop and $\tau' > \tau$, then by the maximality of τ we have $\text{pgen}(\tau') = \emptyset$, contradicting the fact that $\tau' = \wedge\text{pgen}(\tau')$. □

The <u>meet pseudocomplement</u> of $\tau \in$ R-tors, if it exists, is the unique element $\tau^{\perp}$ of R-tors satisfying the conditions

(1) $\tau \wedge \tau^{\perp} = \xi$;

(2) If $\tau' \in$ R-tors and $\tau \wedge \tau' = \xi$ then $\tau' \leq \tau^{\perp}$.

(20.5) PROPOSITION: _If_ $\tau \in$ R-tors _is semiprime then_ $\tau^{\perp}$ _exists and is semiprime. Moreover,_ $\tau^{\perp} = \wedge[\text{R-sp} \setminus \text{pgen}(\tau)]$.

PROOF: Let $\tau' = \wedge[\text{R-sp} \setminus \text{pgen}(\tau)]$. Since τ is semiprime, $\tau \wedge \tau' = \sqrt{\tau} \wedge \tau' = \wedge(\text{R-sp}) = \xi$. On the other hand, if $\tau'' \in$ R-tors satisfies $\tau \wedge \tau'' = \xi$ then for each $\tau_0 \in \text{R-sp} \setminus \text{pgen}(\tau)$ we have $\tau \wedge \tau'' \leq \tau_0$. By Proposition 19.11, this implies that $\tau'' \leq \tau_0$ and so $\tau'' \leq \tau'$. This proves that $\tau' = \tau^{\perp}$. Moreover, this same argument shows that $\text{R-sp} \setminus \text{pgen}(\tau) \subseteq \text{pgen}(\tau^{\perp})$ and so $\tau^{\perp} = \wedge\text{pgen}(\tau^{\perp})$, proving that $\tau^{\perp}$ is semiprime. □

(20.6) PROPOSITION: _If_ $M \in$ R-simp _then_ $\chi(M)^{\perp} > \xi$.

PROOF: If $M \in$ R-simp then $\chi(M)$ is prime and so by Proposition 20.5, $\chi(M)^{\perp}$ exists. Moreover, $\chi(M) \wedge \xi(M) = \xi$ and so $\xi(M) \leq \chi(M)^{\perp}$, proving that $\chi(M)^{\perp} > \xi$. □

We can now extend Proposition 19.10.

(20.7) PROPOSITION: _If_ $\tau \in$ R-sp _and_ $\tau^{\perp} > \xi$ _then_ τ _is a minimal element of_ R-sp.

PROOF: Assume that $\tau \in$ R-sp and that $\tau^{\perp} > \xi$. Suppose further that there exists a $\tau' \in$ R-sp with $\tau' < \tau$. Then $\tau^{\perp} \wedge \tau = \xi < \tau'$ and so by Proposition 19.11 we have $\tau^{\perp} \leq \tau'$. Therefore $\tau^{\perp} = \tau^{\perp} \wedge \tau^{\perp} \leq \tau^{\perp} \wedge \tau' \leq \tau^{\perp} \wedge \tau = \xi$, a contradiction. Therefore τ must be a minimal element of R-sp. □

(20.8) PROPOSITION: <u>The following conditions are equivalent for a ring</u> R:

(1) (i) <u>Every</u> $\tau \in$ R-tors <u>is semiprime</u>.

(ii) <u>For every</u> $U \subseteq$ R-tors,

$$\text{pgen}(\wedge U) = \cup\{\text{pgen}(\tau') \mid \tau' \in U\}.$$

(2) <u>The function</u> $\tau \mapsto \text{pgen}(\tau)$ <u>is a bijection between</u> R-prop <u>and the subsets of</u> R-sp <u>closed under prime generalization</u>.

PROOF: (1) ⇒ (2): If $\tau \in$ R-prop then pgen(τ) is closed under taking prime generalizations. Moreover, by (1i) and the remark after Proposition 20.2, the function $\tau \mapsto \text{pgen}(\tau)$ is monic.

Let $U \subseteq$ R-sp be closed under prime generalization. Then $U \subseteq \text{pgen}(\wedge U)$ and, indeed, by (1ii), they must be equal.

(2) ⇒ (1): If $\tau \in$ R-tors then $\text{pgen}(\tau) = \text{pgen}(\sqrt{\tau})$ and so, by (2), $\tau = \sqrt{\tau}$, proving (1i). If $U \subseteq$ R-tors and $U' = \cup\{\text{pgen}(\tau') \mid \tau' \in U\}$, then by Proposition 8.4, we have $\wedge U' = \wedge\{\wedge\text{pgen}(\tau') \mid \tau' \in U\} = \wedge U$ by (1i). Since U' is

closed under prime generalization, by (2) we have $U' = \mathrm{pgen}(\wedge U')$ and so $\mathrm{pgen}(\wedge U) = \mathrm{pgen}(\wedge U') = U'$, proving (1ii). □

We will say that $\tau \in R\text{-prop}$ is <u>strongly semiprime</u> if and only if $\tau = \wedge\{\chi(M) \mid M \in R\text{-mod}$ is τ-cocritical$\}$. Clearly every strongly semiprime $\tau \in R\text{-prop}$ is semiprime.

(20.9) PROPOSITION: <u>A sufficient condition for $\tau \in R\text{-prop}$ to be strongly semiprime is that every $I \notin \mathcal{L}_\tau$ be contained in a τ-critical left ideal of</u> R.

PROOF: Let $\tau'' = \wedge\{\chi(M) \mid M \in R\text{-mod}$ is τ-cocritical$\}$. Then clearly $\tau'' \geq \tau$. Conversely, let $I \notin \mathcal{L}_\tau$. Then I is contained in a τ-cocritical left ideal H of R. Moreover, $R/H \in \mathcal{F}_{\chi(R/H)} \subseteq \mathcal{F}_{\tau''}$ and so $H \notin \mathcal{L}_{\tau''}$. Hence $I \notin \mathcal{L}_{\tau''}$. Thus we have $\mathcal{L}_{\tau''} \subseteq \mathcal{L}_\tau$ and so $\tau'' \leq \tau$, proving equality. □

If $\tau \in R\text{-fin}$ and $I \notin \mathcal{L}_\tau$, then the τ-purification of I is a proper left ideal of R and so, by Proposition 15.2, is contained in a maximal τ-pure left ideal of R. Therefore I is contained in a maximal τ-pure left ideal of R, which proves that τ is strongly semiprime. In particular, if $\tau \in R\text{-perf}$ then τ is strongly semiprime. Also, by Proposition 12.3, ξ is always strongly semiprime.

(20.10) PROPOSITION: <u>The following conditions are equivalent for</u> $\tau \in R\text{-tors}$:

(1) τ <u>is strongly semiprime</u>.

(2) <u>The family of subsets</u> U <u>of</u> $\mathrm{pgen}(\tau)$ <u>satisfying</u> $\wedge U = \tau$ <u>has a minimal member</u>.

(3) <u>If</u> $I \notin \mathcal{L}_\tau$ <u>there exists a</u> τ-<u>critical left ideal</u> H'' <u>of</u> R <u>and an element</u> $a \in R \setminus I$ <u>such that</u> $(I:a) \subseteq H''$.

(4) $\tau = \wedge U$ <u>for some</u> $U \subseteq \{\chi(M) \mid M \in R\text{-mod}$ <u>is</u> τ-<u>cocritical</u>$\}$.

PROOF: (1) $\Rightarrow$ (2): Let $V = \{\chi(M) \mid M \in R\text{-mod}$ is τ-cocritical$\}$. Then by (1), $\tau = \wedge V$. Therefore, there exists a τ-cocritical left ideal I'. Set $\tau' = \chi(R/I')$. If $\tau = \tau'$ then $\{\tau'\}$ is the minimal set we seek. If $\tau \neq \tau'$ there exists a τ-critical left ideal I'' of R with $\tau'' = \chi(R/I'') \neq \tau'$.

If $I' \notin \mathcal{L}_{\tau''}$ then I' must be τ''-critical, for if H is a left ideal of R properly containing I' then $R/H \in \mathcal{J}_\tau \subseteq \mathcal{J}_{\tau''}$. But by Proposition 19.2, we have $\tau' = \tau''$, which is a contradiction. Therefore, $I' \in \mathcal{L}_{\tau''}$ and so $\wedge(V \setminus \{\tau'\}) > \tau$. Thus V is the minimal set we seek.

(2) $\Rightarrow$ (3): By (2), there exists a subset U of $\mathrm{pgen}(\tau)$ satisfying $\wedge U = \tau$ and minimal with respect to this property. Let $I \notin \mathcal{L}_\tau$. Then there exists a $\tau' \in U$ with $I \notin \mathcal{L}_{\tau'}$, and

so, by Proposition 18.3, there exists a τ'-critical left ideal H of R and an element $a' \in R \smallsetminus I$ with $(I:a') \subseteq H$.

By the minimality of U, there exists a left ideal I' of R such that $I' \notin \mathcal{L}_{\tau'}$ and $I' \in \mathcal{L}_{\tau''}$ for every $\tau'' \in U \smallsetminus \{\tau'\}$. By Proposition 19.3, there exists a $b \in R \smallsetminus I'$ such that $(I':b) \subseteq H'$ for some τ'-critical left ideal H' of R. Since $(I':b) \in \mathcal{L}_{\tau''}$ for all $\tau'' \in U \smallsetminus \{\tau'\}$, we have $H' \in \mathcal{L}_{\tau''}$ for all such τ''. If K is a left ideal of R properly containing H', then $K \in \mathcal{L}_{\tau'}$ by the τ'-criticalness of H' and so $K \in \cap\{\mathcal{L}_{\tau''} \mid \tau'' \in U\} = \mathcal{L}_{\tau}$. Thus H' is τ-critical.

By Proposition 19.5, there exists a $c \in R \smallsetminus H$ and a $c' \in R \smallsetminus H'$ with $(H:c) = (H':c')$. Set $H'' = (H':c')$ and $a = ca'$. Then H' is also τ-critical by Proposition 18.9, and $(I:a) \subseteq (H:c) \subseteq H''$, proving (3).

(3) ⇒ (4): Let $V = \{\chi(M) \mid M \in R\text{-mod}$ is τ-cocritical$\}$. Then $\wedge V \geq \tau$. Conversely, if $I \notin \mathcal{L}_{\tau}$, then by (3) and Proposition 19.3, there exists a $\tau' \in V$ with $I \notin \mathcal{L}_{\tau'}$. Therefore, $\tau = \wedge V$, proving (4). □

(4) ⇒ (1): This proof is trivial.

(20.11) PROPOSITION: <u>For a strongly semiprime</u> $\tau \in R\text{-prop}$ <u>the following conditions are equivalent:</u>

(1) $\tau \in R\text{-sp}$.

(2) $\tau = \tau_1 \wedge \tau_2$ <u>implies that</u> $\tau = \tau_1$ <u>or</u> $\tau = \tau_2$.

(3) $\tau \geq \tau_1 \wedge \tau_2$ _implies that_ $\tau \geq \tau_1$ _or_ $\tau \geq \tau_2$.

PROOF: (1) ⇒ (3) follows from Proposition 19.11 and (3) implies (2) is proven trivially. As for (2) ⇒ (1), let $U \subseteq$ R-tors be minimal with respect to $\tau = \wedge U$. If U is a singleton we are done. If U has more than one member, let $\tau' \in U$ and let $U' = U \setminus \{\tau'\}$. Then $\tau = \tau' \wedge (\wedge U')$ and so by (2) either $\tau = \tau'$ or $\tau = \wedge U'$. By the minimality of U, the second possibility cannot hold and so $\tau = \tau' \in$ R-sp. □

In particular, if every $\tau \in$ R-prop is strongly semiprime, then primeness in the sense of this chapter is the same as lattice-theoretic primeness. By the remark after Proposition 20.9, this occurs when R is left noetherian.

(20.12) PROPOSITION (Raynaud's Theorem): _The following conditions on a ring_ R _are equivalent_:

(1) _Every_ $\tau \in$ R-prop _is strongly semiprime_.

(2) R _is left semiartinian_.

PROOF: (1) ⇒ (2): If $\tau \in$ R-prop then $\chi > \tau = \wedge\{\chi(M) \mid M \in \text{R-mod is } \tau\text{-cocritical}\}$ and hence there must be at least one τ-cocritical left R-module M. Thus R is left seminoetherian.

(2) ⇒ (1): Let $\tau \in$ R-prop and let $\tau' = \wedge\{\chi(M) \mid M \in \text{R-mod is } \tau\text{-cocritical}\}$. If $\tau' > \tau$, then

by Proposition 18.16 there exists a τ-cocritical left R-module that is τ'-torsion, which is a contradiction. Therefore we must have $\tau = \tau'$, which proves (1). □

We now strengthen Proposition 19.17.

(20.13) PROPOSITION: If R is a left semi-noetherian ring then R-sp satisfies the descending chain condition.

PROOF: Let U be a nonempty subset of R-sp and let $\tau = \wedge U$. Set $V = \{\chi(R/I) \mid I$ is a τ-critical left ideal of $R\}$. By Proposition 20.10, we have $\tau = \wedge V$. We claim that $V \subseteq U$. Indeed, let $\tau' = \chi(R/I) \in V$, where I is a τ-critical left ideal of R. Then $I \notin \mathcal{L}_\tau$ and so there exists a $\tau'' \in U$ satisfying $I \notin \mathcal{L}_{\tau''}$. By Proposition 19.3, there exists an $a \in R \setminus I$ such that $(I:a)$ is contained in a τ''-critical left ideal H of R. By Proposition 18.9, $(I:a)$ is again τ-critical. If $(I:a) \subset H$ then $R/H \in \mathcal{T}_\tau \subseteq \mathcal{T}_{\tau''}$ which is a contradiction. Therefore $(I:a) = H$ and so $\tau' = \chi(R/(I:a)) = \tau'' \in U$.

We now claim that the elements of V are minimal elements of U. Indeed, assume that $\tau' \in V$, $\tau'' \in U$, and $\tau'' \leq \tau'$. Then there exists a τ-critical left ideal I of R satisfying $\tau' = \chi(R/I)$. In particular, $R/I \in \mathcal{F}_{\tau'} \subseteq \mathcal{F}_{\tau''}$. On the other hand, if $I' \supset I$ then $R/I' \in \mathcal{T}_\tau \subseteq \mathcal{T}_{\tau''}$.

Therefore I is τ''-critical and so $\tau'' = \chi(R/I) = \tau'$. □

(20.14) PROPOSITION: *The following conditions on a ring* R *are equivalent:*

(1) R *is left seminoetherian and left balanced.*

(2) (i) R *is left balanced;*

(ii) *Every* $\tau \in R\text{-prop}$ *is semiprime;*

(iii) $R\text{-sp}$ *satisfies the descending chain condition.*

(3) (i) R *is left seminoetherian;*

(ii) *Every* $\tau \in R\text{-prop}$ *is clear.*

PROOF: (1) ⇒ (2): Clearly (1) implies (2i) and (2ii). Moreover, (1) implies (2iii) by Proposition 20.13.

(2) ⇒ (3): Let $\tau \in R\text{-prop}$. Since τ is semiprime, $\tau = \sqrt{\tau}$. Let U be the set of minimal elements of $\text{pgen}(\tau)$. (These exist by (2iii).) Then surely $\tau = \wedge U$ so we will be done if we can show that τ is clear.

Let $\tau' \in U$ and let I' be a τ'-critical left ideal of R. Then $R/I' \in \mathcal{T}_{\tau'} \subseteq \mathcal{T}_{\tau}$ and so $I' \notin \mathcal{L}_{\tau}$. Assume that I' is not τ-critical. Then there exists a left ideal H of R strictly containing I' and not belonging to $\mathcal{L}_{\tau}$. Since I' is τ'-critical, we have $H \in \mathcal{L}_{\tau'}$. Therefore there exists a $\tau'' \in U \setminus \{\tau'\}$ with $H \notin \mathcal{L}_{\tau''}$. By Proposition 19.3, there exists an $a \in R \setminus H$ with $(H:a)$ contained in a τ''-critical left ideal I'' of R. Thus $(I':a) \subseteq (H:a) \subseteq I''$. Since R

is left balanced, this implies that $\tau'' \leq \tau'$, contradicting the minimality of τ'. Therefore I' is τ-critical and $\tau' = \chi(R/I')$. This proves (3).

(3) ⇒ (1): This follows from Proposition 19.23. □

References for Section 20

Cahen [24]; Popescu [119, 121]; Raynaud [126, 127]; Schelter [137]; Storrer [149].

21. PRIMARY DECOMPOSITION

Let M be a left R-module. We define the assassin $\mathrm{ass}(M)$ of M by $\mathrm{ass}(M) = \{\tau \in R\text{-sp} \mid$ there exists a τ-cocritical submodule of $M\}$.

(21.1) PROPOSITION: *Let* M *be a left* R-*module. Then*

(1) *If* N *is a submodule of* M, *then* $\mathrm{ass}(N) \subseteq \mathrm{ass}(M) \subseteq \mathrm{ass}(N) \cup \mathrm{ass}(M/N)$.

(2) If $\{M_i\}$ *is a family of submodules of* M *with* $M = \cup M_i$, *then* $\mathrm{ass}(M) = \cup \mathrm{ass}(M_i)$.

(3) *If* $\{M_i\}$ *is a family of submodules of* M *with* $M = \oplus M_i$, *then* $\mathrm{ass}(M) = \cup \mathrm{ass}(M_i)$.

(4) *If* N *is a large submodule of* M, *then* $\mathrm{ass}(N) = \mathrm{ass}(M)$.

(5) *If* M *is* τ-*cocritical for some* $\tau \in R\text{-sp}$,

<u>then</u> $\text{ass}(N) = \{\tau\}$ <u>for every nonzero submodule</u> N <u>of</u> M.

PROOF: (1) Clearly $\text{ass}(N) \subseteq \text{ass}(M)$ for any submodule N of M. Now suppose that $\tau \in \text{ass}(M)$ and let M' be a τ-cocritical submodule of M. If $N \cap M' = 0$ then M' is isomorphic to a submodule of M/N and so $\tau \in \text{ass}(M/N)$. Otherwise, $0 \neq N \cap M'$ is a τ-cocritical submodule of N by Proposition 18.2 and so $\tau \in \text{ass}(N)$.

(2) By (1), $\cup\text{ass}(M_i) \subseteq \text{ass}(M)$. Conversely, if $\tau \in \text{ass}(M)$ and if M' is a τ-cocritical submodule of M, then $M' \cap M_k \neq 0$ for some index k and so $M' \cap M_k$ is a τ-cocritical submodule of M_k. Therefore $\tau \in \cup\text{ass}(M_i)$.

(3) Since M is the union of all finite direct sums of the M_i, then by (2), it suffices to prove the result for finite direct sums only. In fact, it suffices to prove it only for the case of two summands and then to proceed by induction. But the case of two summands follows directly from (1).

(4) By (1), $\text{ass}(N) \subseteq \text{ass}(M)$. Conversely, if $\tau \in \text{ass}(M)$ and if M' is a τ-cocritical submodule of M, then by the largeness of N, we have $0 \neq M' \cap N$. Hence $M' \cap N$ is a τ-cocritical submodule of N and so $\tau \in \text{ass}(N)$.

(5) If M is τ-cocritical, then clearly $\tau \in \text{ass}(N)$ for every nonzero submodule N of M. Conversely, if

$\tau' \in \text{ass}(N)$, then N has a nonzero submodule N' that is both τ'-cocritical and τ-cocritical. By Proposition 19.2, $\tau = \chi(N') = \tau'$. □

(21.2) PROPOSITION: Let M be a left R-module and let $U \subseteq \text{ass}(M)$. Then there exists a submodule N of M satisfying

(1) $\text{ass}(N) = \text{ass}(M) \setminus U$; and

(2) $\text{ass}(M/N) = U$.

PROOF: Let $\mathcal{A} = \{{}_R N' \subseteq M \mid \text{ass}(N') \cap U = \emptyset\}$. Then $\mathcal{A}$ is nonempty since $0 \in \mathcal{A}$. By Proposition 21.1(3), $\mathcal{A}$ is inductive, and so by Zorn's Lemma, $\mathcal{A}$ has a maximal element N. This N clearly satisfies (1).

Let $\tau \in \text{ass}(M/N)$. Then there exists a τ-cocritical submodule M'/N of M/N. By Proposition 21.1(1), $\text{ass}(M') \subseteq \text{ass}(N) \cup \{\tau\}$ and so, by the maximality of N, we have $\tau \in U$. The converse is trivial. □

(21.3) PROPOSITION: If $\tau \in R\text{-prop}$ and $M \in \mathcal{F}_\tau$ then $\text{ass}(M) \subseteq \text{pgen}(\tau)$. The converse holds if every nonzero submodule of M has a nonempty assassin.

PROOF: If $\tau' \in \text{ass}(M)$ and if M' is a τ'-cocritical submodule of M, then $\tau' = \chi(M') \geq \chi(M) \geq \tau$ (since M is τ-torsion-free) and so $\tau' \in \text{pgen}(\tau)$.

Conversely, if every nonzero submodule of M has a nonempty assassin and if M is not τ-torsion-free, then in particular $\mathrm{ass}(T_\tau(M)) \neq \emptyset$. Therefore, there exists a nonzero submodule N of $T_\tau(M)$ with $\chi(N) \in \mathrm{ass}(T_\tau(M)) \subseteq \mathrm{ass}(M) \subseteq \mathrm{pgen}(\tau)$. This implies that $\chi(N) \geq \tau$ and hence that $N \in \mathcal{F}_\tau$, which is a contradiction. □

Let M be a left R-module. We say that M is a D-<u>module</u> if and only if $\mathrm{ass}(M/N) \neq \emptyset$ for every proper submodule N of M (this is trivially satisfied if $M = 0$). For example, semiartinian left R-modules are D-modules, for if M is semiartinian and if N is a proper submodule of M, then M/N has a nonzero simple submodule M' and so $\chi(M') \in \mathrm{ass}(M/N)$.

We say that a ring R is a <u>left</u> D-<u>ring</u> if and only if every left R-module is a D-module or, equivalently, if and only if every nonzero left R-module has a nonempty assassin. Clearly, left semiartinian rings are left D-rings. Moreover, left seminoetherian rings are also left D-rings. Indeed, if R is left seminoetherian and if M is a nonzero left R-module, then there exists a $\chi(M)$-cocritical left R-module and hence, by Proposition 18.2(3), a $\chi(M)$-cocritical submodule of M.

(21.4) PROPOSITION: <u>A ring</u> R <u>is a left</u> D-<u>ring</u>

<u>if and only if</u> ${}_R R$ <u>is a</u> D-<u>module</u>.

PROOF: If R is a left D-ring, then clearly ${}_R R$ is a D-module. Conversely, assume that ${}_R R$ is a D-module and let M be a nonzero left R-module. If $0 \neq m \in M$, then Rm is isomorphic to a homomorphic image of ${}_R R$ and so $\text{ass}(Rm) \neq \emptyset$. Therefore, $\text{ass}(M) \neq \emptyset$. □

(21.5) PROPOSITION: <u>The following conditions are equivalent for a ring</u> R:

(1) R <u>is left semiartinian</u>.

(2) (i) R <u>is a left</u> D-<u>ring; and</u>

(ii) <u>Every</u> $\tau \in$ R-sp <u>has a simple</u> τ-<u>cocritical module</u>.

PROOF: (1) ⇒ (2): We have already observed that (1) implies (2i). Moreover, (2ii) follows from (1) by Proposition 19.15.

(2) ⇒ (1): Let M be a nonzero left R-module. By (2i), $\text{ass}(M) \neq \emptyset$ and so M has a cocritical submodule N. By (2ii), $\chi(N)$ has a simple τ-cocritical module N'. By Proposition 18.2, there is a submodule of N isomorphic to N'. Therefore, M has a simple submodule, proving (1). □

(21.6) PROPOSITION (Theorem of Akizuki for Noncommutative Rings): <u>The following conditions are equivalent for a ring</u> R:

(1) R *is left artinian.*

(2) (i) R *is left noetherian; and*

(ii) *Every* $\tau \in R\text{-sp}$ *has a simple* τ-*cocritical module*.

PROOF: (1) ⇒ (2) follows from Proposition 21.5 and the well-known fact that every left artinian ring is left noetherian. (2) ⇒ (1) follows from Propositions 21.5 and 12.8. □

(21.7) PROPOSITION: *Let* $\tau \in R\text{-tors}$. *Then the following conditions are equivalent for a* D-module M:

(1) M *is* τ-*torsion*.

(2) $\text{supp}(M) \cap \text{pgen}(\tau) = \emptyset$.

PROOF: If M is τ-torsion, then M is τ'-torsion for every $\tau' \in \text{pgen}(\tau)$ and so, clearly, $\text{supp}(M) \cap \text{pgen}(\tau) = \emptyset$. Conversely, assume that M is not τ-torsion. Then $0 \neq M/T_\tau(M)$ and so there exists a $\tau' \in \text{ass}(M/T_\tau(M))$. Since $M/T_\tau(M)$ is τ-torsion-free, $\tau \leq \tau'$ whence $\tau' \in \text{pgen}(\tau)$. Moreover, $M/T_\tau(M) \notin \mathcal{J}_{\tau'}$ and so $M \notin \mathcal{J}_{\tau'}$. Therefore, $\tau' \in \text{supp}(M)$. □

(21.8) PROPOSITION: *If* M *is a* D-*module, then* $\text{ass}(N) \neq \emptyset$ *for every nonzero submodule* N *of* M.

PROOF: Let N be a nonzero submodule of M. If

N is large in M, then $\text{ass}(N) = \text{ass}(M) \neq \emptyset$ by Proposition 21.1. If N is not large in M, then $\{{}_R N' \subseteq M \mid N \cap N' = 0\}$ is nonempty and, by Zorn's Lemma, contains a maximal member N'. Then the R-homomorphism defined by the composition of the canonical homomorphisms $N \to M \to M/N'$ is a monomorphism the image of which is large in M/N'. Therefore $\text{ass}(N) = \text{ass}(M/N') \neq \emptyset$. □

(21.9) COROLLARY: <u>If</u> M <u>is a</u> D-<u>module, then the</u> R-<u>homomorphism</u> $\psi: M \to \Pi\{Q_\tau(M) \mid \tau \in \text{ass}(M)\}$ <u>defined by</u> $m \longmapsto \langle m\hat{\tau}_M \rangle$ <u>is a monomorphism</u>.

PROOF: If ψ were not a monomorphism, then $\ker(\psi)$ would be a nonzero submodule of M with empty assassin, contradicting Proposition 21.8. □

(21.10) PROPOSITION: <u>There exists a torsion theory</u> $\tau_D \in$ R-tors <u>with</u> $\mathcal{J}_{\tau_D} = \{M \in \text{R-mod} \mid M$ <u>is a</u> D-<u>module</u>$\}$.

PROOF: We must show that the class of all D-modules satisfies the conditions listed in Proposition 1.6. Let M be a nonzero D-module and let N be a proper submodule of M. If N'/N is a proper submodule of M/N, then $(M/N)/(N'/N) \cong M/N'$, which has a nonempty assassin. Therefore, M/N is also a D-module. If N' is a proper submodule of N, then N/N' is a nonzero submodule of M/N'

and hence has a nonempty assassin by Proposition 21.8. Therefore N is also a D-module.

Now assume that N is a proper submodule of a left R-module M and that both N and M/N are D-modules. Let M' be a proper submodule of M. If $N \subseteq M'$ then M/M' is an epimorphic image of M/N and hence has a nonempty assassin since M/N is a D-module. If $N \not\subseteq M'$ then $N/[M' \cap N] \cong [M' + N]/M' \subseteq M/M'$. Since N is a D-module, $N/[M' \cap N]$ has a nonempty assassin and so M/M' has a nonempty assassin. Therefore M is a D-module.

Finally, we have to show that a direct sum of D-modules is again a D-module. If $M = \oplus M_i$ where each M_i is a D-module and if N is a proper submodule of M, then there exists an index i for which $M_i \not\subseteq N$. Therefore $M_i/[M_i \cap N] \neq 0$. Since M_i is a D-module, $\mathrm{ass}(M_i/[M_i \cap N]) \neq \emptyset$. Furthermore, $M_i/[M_i \cap N] \cong [M_i + N]/N$, which is a submodule of M/N and so $\mathrm{ass}(M/N) \neq \emptyset$. □

(21.11) PROPOSITION: <u>If</u> $\tau \in R\text{-tors}$ <u>and</u> $\tau \vee \tau_D = \chi$, <u>then</u> τ <u>is semiprime</u>.

PROOF: Assume that $M \in \mathcal{T}_{\sqrt{\tau}} \setminus \mathcal{T}_\tau$. Then $\bar{M} = M/T_\tau(M)$ is τ-torsion-free. Since $\mathcal{F}_\tau \cap \mathcal{F}_{\tau_D} = \mathcal{F}_{\tau \vee \tau_D} = \mathcal{F}_\chi = \{0\}$, we have $T_{\tau_D}(\bar{M}) \neq 0$ and so $\bar{M}$ has a nonzero D-submodule N. Pick $\tau' \in \mathrm{ass}(N)$. Then $\mathcal{F}_{\tau'} \subseteq \mathcal{F}_\tau$ and so $\tau' \in \mathrm{pgen}(\tau)$ and $N \notin \mathcal{T}_{\tau'}$. But $M \in \mathcal{T}_{\sqrt{\tau}} \Rightarrow \bar{M} \in \mathcal{T}_{\sqrt{\tau}} \Rightarrow N \in \mathcal{T}_{\sqrt{\tau}} \Rightarrow N \in \mathcal{T}_{\tau''}$ for all

$\tau'' \in \mathrm{pgen}(\tau) \Rightarrow N \in \mathcal{J}_{\tau'}$, which is a contradiction. Therefore, $\sqrt{\tau} = \tau$ and so τ is semiprime. □

(21.12) COROLLARY: _If_ R _is a left_ D-_ring, then every_ $\tau \in$ R-prop _is semiprime_.

PROOF: If R is a left D-ring, then $\tau_D = \chi$ and so the result follows immediately from Proposition 21.11. □

The following result relates prime torsion theories and isomorphism classes of indecomposable injective modules, extending the well-known result for commutative rings.

(21.13) PROPOSITION: _For a ring_ R _there is an injection of_ R-sp _into the collection of all isomorphism classes of indecomposable injective left_ R-_modules_. _If_ R _is a left_ D-_ring, then this is in fact a bijection_.

PROOF: Let $\tau \in$ R-sp and let M be a τ-cocritical left R-module. By Proposition 18.2, M is uniform and so E(M) is indecomposable. If M' is any other τ-cocritical left R-module, then E(M') is also indecomposable. By Proposition 19.6, $E_\tau(M)$ and $E_\tau(M')$ are isomorphic and so E(M) and E(M') are isomorphic. Therefore, the function $\tau \mapsto \{E \in R\text{-mod} \mid E \cong E(M)$ for some τ-cocritical left R-module $M\}$ is well defined; it is an injection by Proposition 19.2.

If R is a left D-ring and if E is an indecomposable injective left R-module, then by Proposition 21.8, E has a nonzero cocritical submodule M and E(M) = E by indecomposability. Therefore, $\tau = \chi(M)$ is prime and E is in the image of τ under the above function. □

(21.14) PROPOSITION: *Let* R *be a left noetherian ring and let* M *be a left* R-*module for which every* $\tau \in \text{supp}(M)$ *is stable. Then* $\text{supp}(M) = \cup\{\text{spcl}(\tau) \cap R\text{-sp} \mid \tau \in \text{ass}(M)\}$.

PROOF: If $\tau \in \text{ass}(M)$, there exists a τ-cocritical submodule N of M. If $\tau' \in \text{spcl}(\tau) \cap R\text{-sp}$, then $\mathcal{F}_\tau \subseteq \mathcal{F}_{\tau'}$ and so N is τ'-torsion-free. Therefore, $M \notin \mathcal{T}_{\tau'}$ and so $\tau' \in \text{supp}(M)$. Conversely, if $\tau' \in \text{supp}(M)$ then $M \notin \mathcal{T}_{\tau'}$. Since R is left noetherian, we can write E(M) as $\oplus E_i$ where the E_i are indecomposable injective left R-modules. Since M is not τ'-torsion, there exists an index i with $E_i \notin \mathcal{T}_{\tau'}$. By Proposition 11.3, E_i is τ'-torsion-free. Therefore, $\tau' \leq \chi(E_i)$ and $\chi(E_i) \in \text{ass}(E(M)) = \text{ass}(M)$ by Proposition 21.13. □

The following result shows that we can add yet another property to the list in Proposition 17.13.

(21.15) PROPOSITION: *If* $\tau \in R\text{-perf}$ *and* R *is a left* D-*ring, then* R_τ *is a left* D-*ring*.

PROOF: Let N be a nonzero left R_τ-module. Since R is a left D-ring, ${}_RN$ has a cocritical R-submodule N'. Since τ is perfect, N is absolutely τ-pure as a left R-module, and so $\chi(N') \in \mathrm{pgen}(\tau)$. Moreover, $N' \subseteq Q_\tau(N') \subseteq N$. By Proposition 18.3, $Q_\tau(N')$ is also $\chi(N')$-cocritical. If $\sigma = \hat{\tau}_\#(\chi(N'))$, then by Proposition 18.4, $Q_\tau(N')$ is σ-cocritical. Therefore by Proposition 19.24, $\sigma \in R_\tau$-sp and so $\sigma \in \mathrm{ass}(N)$. □

(21.16) PROPOSITION: *If I is a two-sided ideal of a left D-ring R, then R/I is a left D-ring.*

PROOF: Let $\gamma: R \to R/I$ be the canonical ring surjection. If N is a nonzero left R/I-module, then ${}_RN$ is a nonzero left R-module and so has a cocritical submodule N'. Since $IN = 0$, we have $IN' = 0$ and so N' is also a left R/I-submodule of N and N' is $\chi(N')$-cocritical as a left R/I-submodule. Therefore, $\chi(N') \in \mathrm{ass}(N)$. □

We now characterize those left R-modules satisfying the condition that every nonzero submodule has a nonempty assassin. (As we have seen in Proposition 21.8, D-modules satisfy this condition.)

(21.17) PROPOSITION: *The following conditions are equivalent for a left R-module M:*

(1) *For every nonzero submodule N of M,*

$\mathrm{ass}(N) \neq \emptyset$.

(2) M _has a large submodule of the form_ $\oplus R/I_j$ _where the_ I_j _are critical left ideals of_ R.

(3) $E(M)$ _has a large submodule of the form_ $\oplus E(R/I_j)$ _where the_ I_j _are critical left ideals of_ R.

PROOF: (1) $\Rightarrow$ (2): Since $\mathrm{ass}(M) \neq \emptyset$, there exists an $m \in M$ with $(0:m)$ critical. Therefore, $\mathcal{A} = \{\oplus R/I_j \subseteq M \mid I_j$ is a critical left ideal of $R\}$ is nonempty. Clearly, $\mathcal{A}$ is inductive and so, by Zorn's Lemma, it contains a maximal member $N = \oplus R/I_j$. We claim that N is large in M. If it is not, there exists a nonzero submodule N' of M with $N \cap N' = 0$. By (1), $\mathrm{ass}(N') \neq \emptyset$ and so there exists an $x \in N'$ with $(0:x)$ critical. Then $N \oplus Rx \in \mathcal{A}$, contradicting the maximality of N. Thus N is large in M.

(2) $\Rightarrow$ (3): If $\oplus R/I_j$ is large in M, then it is large in $E(M)$ and hence $\oplus E(R/I_j)$ is large in $E(M)$.

(3) $\Rightarrow$ (1): Let N be a nonzero submodule of M and let $\{I_j\}$ be a family of critical left ideals of R such that $\oplus E(R/I_j)$ is large in $E(M)$. Then $N \cap [\oplus R/I_j] \neq 0$. Let $0 \neq x \in N \cap [\oplus R/I_j]$. Then $x = \sum_{i=1}^{n} x_i$ for $x_i \in R/I_{j_i}$. We claim that $\mathrm{ass}(Rx) \neq \emptyset$. To show this, it suffices to consider the case of $n = 2$ and from there to proceed by

induction. Therefore, assume that $Rx \subseteq R/I_1 \oplus R/I_2$. If $Rx \cap R/I_1 \neq 0$, then $\chi(R/I_1) \in \text{ass}(Rx)$. If $Rx \cap R/I_1 = 0$, then the canonical R-homomorphism $Rx \to R/I_1 \oplus R/I_2 \to R/I_2$ is a monomorphism and so Rx is isomorphic to a submodule of R/I_2, whence $\chi(R/I_2) \in \text{ass}(Rx)$. Thus $\text{ass}(Rx) \neq \emptyset$ and so $\text{ass}(N) \neq \emptyset$. □

A left R-module M is called <u>coprimary</u> if and only if $\text{ass}(M)$ consists of precisely one member.

(21.18) PROPOSITION: *Let R be a left noetherian ring. Then for a left R-module M the following conditions are equivalent:*

(1) *M is coprimary.*

(2) *There exists an indecomposable injective left R-module E with $E(M) \cong E^{(\Omega)}$ for some index set Ω.*

PROOF: (1) ⇒ (2): Let M be a coprimary left R-module with $\text{ass}(M) = \{\tau\}$. Then by Proposition 21.13, $\tau = \chi(E)$ for some indecomposable injective left R-module E. If E' is an indecomposable injective submodule of $E(M)$, then $\chi(E') \in \text{ass}(M)$ and so $\chi(E') = \tau$. By Proposition 21.13 we therefore have $E \cong E'$. Since R is left noetherian, $E(M) = \oplus\{E_i \mid i \in \Omega\}$ for some indecomposable injective left R-modules E_i and index set Ω. By the above each E_i is isomorphic to E and so we have (2).

(2) ⇒ (1): By (2), there exists indecomposable injective left R-module E such that $E(M) = \oplus\{E_i \mid i \in \Omega\}$ where each E_i is isomorphic to E. Then $\tau = \chi(E)$ is clearly a member of $\mathrm{ass}(M)$.

Assume that $\tau' \in \mathrm{ass}(M)$ and let $m \in M$ satisfy $\tau' = \chi(Rm)$. Then there exists a finite subset Ω' of Ω with $m \in \oplus\{E_i \mid i \in \Omega'\}$. Hence $0 \neq \mathrm{Hom}_R(Rm, \oplus_{i\in\Omega'} E_i) \cong \oplus_{i\in\Omega'} \mathrm{Hom}_R(Rm, E_i)$ and so in particular $\mathrm{Hom}_R(Rm,E) \neq 0$. By Proposition 18.2, we have an embedding $Rm \to E$ and so, by the indecomposability of E, we have $E(Rm) \cong E$. Therefore, $\tau' = \chi(Rm) = \tau$ and so M is comprimary. □

(21.19) PROPOSITION: *If R is left semiartinian, then the following conditions are equivalent for a nonzero left R-module M:*

(1) *M is coprimary.*

(2) *Any two simple submodules of M are isomorphic.*

PROOF: (1) ⇒ (2): If M' is a simple submodule of M, then M' is cocritical and so $\tau' = \chi(M')$ is prime. Therefore, $\tau' \in \mathrm{ass}(M)$. If M'' is another simple submodule of M, then by (1) we must have $\chi(M'') = \tau'$. Therefore, there exists a nonzero R-homomorphism $M'' \to E(M')$ which, by the simplicity of M' and M'', is an isomorphism $M'' \to M'$.

(2) ⇒ (1): Since R is left semiartinian, M has at least one simple submodule. Moreover, for every simple submodule M' of M, $\chi(M') \in \text{ass}(M)$. Thus $\text{ass}(M) \neq \emptyset$. If $\tau \in \text{ass}(M)$, then there exists a τ-cocritical submodule N of M. Since $0 \neq N$, the module N has a simple submodule N' which is also τ-cocritical by Proposition 18.2. By Proposition 19.2, $\tau = \chi(N')$. Therefore, every $\tau \in \text{ass}(M)$ is of the form $\chi(M')$ for some simple submodule M' of M. It is now immediate that (2) implies (1). □

(21.20) PROPOSITION: <u>A nonzero uniform noetherian left R-module is coprimary</u>.

PROOF: Let M be a nonzero uniform noetherian left R-module and set $\tau = \chi(M)$. Let $\mathcal{A}$ be the family of all proper submodules of M that are not τ-dense in M. Then $\mathcal{A}$ is nonempty since it contains 0. Since M is noetherian, $\mathcal{A}$ has a maximal element N. Since $M/N \notin \mathcal{J}_\tau$, there is a nonzero R-homomorphism $\alpha: M/N \to E(M)$. Since $(M/N)/\ker(\alpha)$ is τ-torsion-free, we must have $\ker(\alpha) = 0$, and so N is τ-pure in M. Therefore M/N is τ-cocritical.

Since M is large in E(M), $0 \neq M' = M \cap (M/N)\alpha$ and so M' is a τ-cocritical submodule of M. Therefore, $\chi(M') \in \text{ass}(M)$ and so $\text{ass}(M) \neq \emptyset$. Now assume that $\tau' \in \text{ass}(M)$ and let M" be a τ'-cocritical submodule of M. By the uniformity of M, we have $0 \neq M' \cap M''$. By

Proposition 21.1(4), $\{\tau'\} = \text{ass}(M' \cap M'') = \{\chi(M')\}$.
Therefore M is coprimary. □

(21.21) PROPOSITION: Every nonzero noetherian left R-module is a D-module.

PROOF: If M is noetherian, then so is M/N for every proper submodule N of M and so it suffices to show that for every nonzero noetherian left R-module M, $\text{ass}(M) \neq \emptyset$. By Proposition 21.20, it suffices to show that every nonzero noetherian left R-module contains a nonzero uniform submodule.

Let M be a nonzero left R-module. If M is not uniform, then there exist nonzero submodules N_1 and N_1' of M such that $N_1 \cap N_1' = 0$. If N_1' is not uniform, then there exist nonzero submodules N_2 and N_2' of N_1' with $N_2 \cap N_2' = 0$. Continue in this manner. If we do not obtain a nonzero uniform submodule of M in a finite number of steps then we obtain an infinite ascending chain $N_1 \subset N_1 \oplus N_2 \subset \cdots$, contradicting the fact that M is noetherian. □

(21.22) PROPOSITION: If M is a nonzero noetherian left R-module, then $\text{ass}(M)$ is finite.

PROOF: Since M is noetherian, every submodule of M is a finite reduced intersection of meet-irreducible submodules, by a well-known result (see, for example, [166], Volume I, p. 208). Thus, in particular, $0 = \bigcap_{i=1}^{n} N_i$ where the N_i are meet irreducible. Then M/N_i is uniform and noetherian for each i and so, by Proposition 21.20, each M/N_i is coprimary. Let $\text{ass}(M/N_i) = \{\tau_i\}$. We have a canonical R-monomorphism $M \to \oplus M/N_i$ and so, by Proposition 21.1, $\text{ass}(M) \subseteq \text{ass}(\oplus M/N_i) = \cup\text{ass}(M/N_i) = \{\tau_1,\ldots,\tau_n\}$. Since $\text{ass}(M) \neq \emptyset$ by Proposition 21.21, we conclude that $\text{ass}(M)$ is finite. □

A submodule N of a left R-module M is <u>primary</u> in M if and only if M/N is coprimary. By Proposition 21.1(5), critical left ideals of R are primary submodules of R. In particular, if $\text{ass}(M/N) = \{\tau\}$ we say that N is τ-<u>primary</u> in M.

(21.23) PROPOSITION: *If $\{N_1,\ldots,N_k\}$ is a finite set of τ-primary submodules of a D-module M, then $N = \cap N_i$ is also a τ-primary submodule of M.*

PROOF: Clearly, M/N is isomorphic to a nonzero submodule of $\oplus M/N_i$. By Proposition 21.1, $\text{ass}(\oplus M/N_i) = \cup\text{ass}(M/N_i) = \{\tau\}$. Since M is a D-module, $\text{ass}(M/N) \neq \emptyset$ and so $\text{ass}(M/N) = \{\tau\}$. □

A submodule N of a left R-module M is said to have a <u>primary decomposition</u> if and only if there exists a nonempty family $\{N_i \mid i \in \Omega\}$ of submodules of M satisfying the following conditions:

(1) Each N_i is primary in M.

(2) $N = \cap\{N_i \mid i \in \Omega\}$ and for any proper subset Ω' of Ω, we have $N \neq \cap\{N_i \mid i \in \Omega'\}$.

(3) $\text{ass}(M/N_i) \neq \text{ass}(M/N_j)$ for all $i \neq j \in \Omega$.

(4) $\text{ass}(N_i/N) = \text{ass}(M/N) \setminus \text{ass}(M/N_i)$ for all $i \in \Omega$.

(5) $\text{ass}(M/N) = \cup\{\text{ass}(M/N_i) \mid i \in \Omega\}$.

(21.24) PROPOSITION: A left R-module M is a D-module if and only if every proper submodule of M has a primary decomposition.

PROOF: Let M be a D-module and let N be a proper submodule of M. By Proposition 21.10, M/N is also a D-module and so, replacing M by M/N if necessary, it suffices to assume that $N = 0$. For each $\tau \in \text{ass}(M)$ there exists, by Proposition 21.2, a submodule N_τ of M with $\text{ass}(M/N_\tau) = \{\tau\}$ and $\text{ass}(N_\tau) = \text{ass}(M) \setminus \{\tau\}$. Then N_τ is a primary submodule of M. Furthermore, $\{N_\tau \mid \tau \in \text{ass}(M)\}$ also clearly satisfies conditions (3), (4), and (5) of the definition of a primary decomposition. If $N' = \cap\{N_\tau \mid \tau \in \text{ass}(M)\}$, then $\text{ass}(N') \subseteq \text{ass}(N_\tau)$ for each $\tau \in \text{ass}(M)$ and so $\text{ass}(N') = \emptyset$. By Proposition 21.8, this implies that $N' = 0$. Moreover, if Ω' is a proper subset of ass(M) and if $N'' = \cap\{N_\tau \mid \tau \in \Omega\}$, then

$\mathrm{ass}(N'') = \mathrm{ass}(M) \setminus \Omega' \neq \emptyset$ and so $N'' \neq 0$. Therefore, $\{N_\tau \mid \tau \in \mathrm{ass}(M/N)\}$ yields a primary decomposition of N.

Conversely, suppose that every proper submodule of M has a primary decomposition, and let N be a proper submodule of M. Then in particular N is contained in a primary submodule N' of M. Then by condition (5) of the definition of a decomposition, $\mathrm{ass}(M/N) \supseteq \mathrm{ass}(M/N') \neq \emptyset$. Therefore, M is a D-module. □

(21.25) COROLLARY: <u>Every left</u> R-<u>module contains a unique submodule maximal with respect to the property that every proper submodule of it has a primary decomposition</u>.

PROOF: This follows from Propositions 21.10 and 21.24. □

(21.26) COROLLARY (Goldman Primary Decomposition Theorem): <u>If</u> M <u>is a nonzero noetherian left</u> R-<u>module, then every proper submodule of</u> M <u>has a finite primary decomposition</u>.

PROOF: This follows from Propositions 21.24, 21.21, and 21.22. □

(21.27) PROPOSITION: <u>If</u> $\tau \in$ R-perf, <u>then for every left</u> R_τ-<u>module</u>, N, <u>we have</u> $\hat{\tau}^{\#}(\mathrm{ass}(N)) \subseteq \mathrm{ass}({}_RN)$. <u>If</u> R <u>is left noetherian</u>

then we have equality.

PROOF: Let $\sigma' \in \mathrm{ass}(N)$. By Corollary 19.25, $\tau' = \hat{\tau}^{\#}(\sigma') \in R\text{-sp}$. If N' is a σ'-cocritical submodule of N, then ${}_RN'$ is τ'-cocritical by Proposition 18.4 and so $\tau' \in \mathrm{ass}({}_RN)$.

Conversely, let $\tau' = \chi(R/I) \in \mathrm{ass}({}_RN)$ where $I = (0:x)$ for some $x \in N$. Set $H = \{s \in R_\tau \mid sx = 0\}$. By Proposition 17.13, R_τ is left noetherian if R is, and so by Corollary 21.26, we can write $H = \bigcap_{j=1}^{n} K_j$ where the K_j and left ideals of R_τ with $\mathrm{ass}(R_\tau/K_j) = \{\sigma_j\}$ for $j = 1,\ldots,n$. Furthermore, $\sigma' = \chi(R_\tau x)$ satisfies $\sigma' = \wedge\sigma_j$ and $\mathrm{ass}(R_\tau x) = \{\sigma_1,\ldots,\sigma_n\} \subseteq \mathrm{ass}(N)$.

For each j, with $1 \leq j \leq n$, set $\tau_j = \hat{\tau}^{\#}(\sigma_j)$. We claim that $\tau' = \wedge\tau_j$. Indeed, $\sigma' \leq \sigma_j$ implies that $\tau' \leq \tau_j$ for each j and so $\tau' \leq \wedge\tau_j$. Now assume that the inequality is strict. Then there exists a nonzero left R-module M belonging to $[\cap\mathcal{T}_{\tau_j}] \setminus \mathcal{T}_{\tau'}$. Replacing M by $M/T_{\tau'}(M)$ if necessary, we can in fact assume that M is τ'-torsion-free. Then $Q_\tau(M) \cong R_\tau \otimes_R M \in \mathcal{F}_{\sigma'}$. On the other hand, $M \in \mathcal{T}_{\tau_j}$ implies that $T_{\tau_j}(Q_\tau(M))$ is large in $Q_\tau(M)$. Therefore, since this is true for all j, $T_{\sigma'}(Q_\tau(M)) = \cap T_{\sigma_j}(Q_\tau(M)) = \cap T_{\tau_j}(Q_\tau(M))$ is large in $Q_\tau(M)$, which is a contradiction. Thus $\tau' = \wedge\tau_j$. Since τ' is prime, by Proposition 19.11 we have $\tau' = \tau_j$ for some j and so $\tau' \in \hat{\tau}^{\#}(\mathrm{ass}(N))$. □

References for Section 21

Cahen [24]; Goldman [63]; Hudry [75, 76, 77]; Michler [101]; Nastasescu [110]; Nguyen-Trong-Kham [114, 115, 116]; Popescu [119, 120, 121]; Raynaud [126]; Storrer [149].

CHAPTER V

DECOMPOSITION OF TORSION THEORIES

22. JANSIAN TORSION THEORIES

If $\tau \in$ R-tors, we recall that $L(\tau)$, the intersection of all τ-dense left ideals of R, is a two-sided ideal of R which, in general, does not belong to $\mathcal{L}_\tau$. If $L(\tau) \in \mathcal{L}_\tau$, we say that τ is *jansian*. We denote the family of all jansian $\tau \in$ R-tors by R-jans.

(22.1) PROPOSITION: *The function* $\tau \mapsto L(\tau)$ *is a bijective correspondence between* R-jans *and the family of all idempotent two-sided ideals of* R.

PROOF: Suppose τ is jansian. We know that $L(\tau)$ is a two-sided ideal of R and clearly $L(\tau)^2 \subseteq L(\tau)$. Moreover, if $a \in L(\tau)$ then $(L(\tau)^2:a) \supseteq L(\tau)$ and so $(L(\tau)^2:a) \in \mathcal{L}_\tau$. Since this is true for all $a \in L(\tau) \in \mathcal{L}_\tau$, we have $L(\tau)^2 \in \mathcal{L}_\tau$ and so $L(\tau)^2 = L(\tau)$ by the minimality of $L(\tau)$. This proves that $L(\tau)$ is idempotent.

Conversely, let I be an idempotent two-sided ideal of R. Then it is easy to show that $\{M \in \text{R-mod} \mid IM = 0\}$

satisfies the conditions in Proposition 1.6(2) and so it is $\mathcal{T}_\tau$ for some $\tau \in$ R-tors. Then $\mathcal{L}_\tau = \{{}_RH \subseteq R \mid I(R/H) = 0\} = \{{}_RH \subseteq R \mid I \subseteq H\}$ and so $I = L(\tau) \in \mathcal{L}_\tau$. Thus τ is jansian.

Finally, to show bijectiveness we must show that if $\tau \in$ R-jans and $M \in \mathcal{T}_\tau$, then $L(\tau)M = 0$. Assume that this is not so. Then there exists an $m \in M$ with $L(\tau)m \neq 0$ and so there exists a left ideal H of R, contained in $L(\tau)$, with $L(\tau)/H \cong L(\tau)m \in \mathcal{T}_\tau$. By the exactness of the sequence $0 \to L(\tau)/H \to R/H \to R/L(\tau) \to 0$, it follows that $H \in \mathcal{L}_\tau$ whence $H \supseteq L(\tau)$, which is a contradiction. Therefore $L(\tau)M = 0$. □

We therefore note, in particular, that if $\tau \in$ R-jans, then a left R-module M is τ-torsion if and only if every element of M is annihilated by $L(\tau)$, i.e., if and only if M is canonically an $R/L(\tau)$-module.

(22.2) PROPOSITION: <u>The following conditions are equivalent for</u> $\tau \in$ R-tors:

(1) τ <u>is jansian</u>.

(2) $L(\tau)$ <u>is idempotent and</u>

$\mathcal{T}_\tau \supseteq \{M \in R\text{-mod} \mid \mathrm{Hom}_R(L(\tau),M) = 0\}$.

(3) $\mathcal{T}_\tau = \{M \in R\text{-mod} \mid \mathrm{Hom}_R(L(\tau),M) = 0\}$.

(4) $\mathcal{T}_\tau$ <u>is closed under taking direct products</u>.

PROOF: (1) ⇒ (2): By Proposition 22.1, $L(\tau)$ is idempotent. In addition, if $M \notin \mathcal{T}_\tau$, there exists an

$0 \neq m \in M$ such that $L(\tau)m \neq 0$ and so m defines a nonzero R-homomorphism $L(\tau) \to M$ given by $a \mapsto am$.

(2) ⇒ (3): Let $M \in \mathcal{T}_\tau$. Since $L(\tau)$ is idempotent, $\mathrm{Hom}_R(L(\tau),R/L(\tau)) = 0$ and so $R/L(\tau) \in \mathcal{T}_\tau$ by (2). If $\alpha \in \mathrm{Hom}_R(L(\tau),M)$, then $L(\tau)/\ker(\alpha) \in \mathcal{T}_\tau$ and so, from the exactness of the sequence $0 \to L(\tau)/\ker(\alpha) \to R/\ker(\alpha) \to R/L(\tau) \to 0$, we obtain that $R/\ker(\alpha) \in \mathcal{T}_\tau$, i.e., $\ker(\alpha) \supseteq L(\tau)$. Therefore $\alpha = 0$. Together with (2), this proves (3).

(3) ⇒ (4): If $\{M_i\}$ is a family of τ-torsion left R-modules, then $\mathrm{Hom}_R(L(\tau),\Pi M_i) = \Pi\mathrm{Hom}_R(L(\tau),M_i) = 0$ and so $\Pi M_i \in \mathcal{T}_\tau$ by (3).

(4) ⇒ (1): Let $M = \Pi\{R/I \mid I \in \mathcal{L}_\tau\}$. By (4), M is τ-torsion. Let $\alpha: R \to M$ be the canonical R-homomorphism $r \mapsto \langle r+I \rangle$. Then $R/\ker(\alpha)$ is τ-torsion and so $\ker(\alpha) \in \mathcal{L}_\tau$. But $\ker(\alpha) \subseteq I$ for every $I \in \mathcal{L}_\tau$ and so we must have $\ker(\alpha) = L(\tau)$, proving that τ is jansian. □

Let $\tau \in$ R-tors. In Proposition 5.5, we showed that every τ-torsion-free left R-module is absolutely τ-pure if and only if there exists a $\tau' \in$ R-tors with $\mathcal{F}_\tau = \mathcal{T}_{\tau'}$. Then $\mathcal{T}_{\tau'}$ is closed under taking direct products, and so by Proposition 22.2, we see that in fact $\tau' \in$ R-jans.

(22.3) PROPOSITION: <u>If</u> $\tau \in$ R-jans, <u>then for any</u>

<u>family</u> $\{M_i\}$ <u>of left</u> R-<u>modules</u>, $T_\tau(\Pi M_i) = \Pi T_\tau(M_i)$.

PROOF: By Proposition 22.2, $\Pi T_\tau(M_i)$ is τ-torsion and so $\Pi T_\tau(M_i) \subseteq T_\tau(\Pi M_i)$. Conversely, if $x = \langle x_i \rangle \in T_\tau(\Pi M_i)$ then each x_i belongs to the homomorphic image of $T_\tau(\Pi M_i)$ under the canonical R-epimorphism $\Pi M_i \to M_i$ and so $x \in \Pi T_\tau(M_i)$, proving equality. □

(22.4) PROPOSITION: R-jans <u>is closed under taking meets</u>.

PROOF: Let $U \subseteq$ R-jans. If $\{M_i\}$ is a family of $\wedge U$-torsion modules, then $\{M_i\} \subseteq \mathcal{J}_\tau$ for each $\tau \in U$. By Proposition 22.2, $\Pi M_i \in \mathcal{J}_\tau$ for each $\tau \in U$ and so $\Pi M_i \in \mathcal{J}_{\wedge U}$. Again by Proposition 22.2, this shows that $\wedge U \in$ R-jans. □

(22.5) PROPOSITION: <u>A ring</u> R <u>is left semiartinian if and only if</u> ξ(R-simp) <u>is stable and jansian</u>.

PROOF: If R is left semiartinian, then ξ(R-simp) = χ, which is clearly stable and jansian. Conversely, assume that ξ(R-simp) is stable and jansian. Then for any simple left R-module M, we have $E(M) \in \mathcal{J}_{\xi(R\text{-simp})}$. By Proposition 22.2, any direct product of injective hulls of simple left R-modules is ξ(R-simp)-torsion. But it is well known that any left R-module can be embedded in a direct product of injective hulls of simple left R-modules and thus $\xi(\text{R-simp}) = \chi$, proving that R is left semiartinian. □

(22.6) PROPOSITION: *If* $\tau \in$ R-jans, *then for any left* R-*module* M, *we have* $Q_\tau(M) \cong \mathrm{Hom}_R(L(\tau), M/T_\tau(M))$.

PROOF: Let M be a left R-module and let $\bar{M} = M/T_\tau(M)$. If $\alpha\colon L(\tau) \to \bar{M}$ is an R-homomorphism, then by τ-injectivity there exists a unique R-homomorphism β making the diagram

$$\begin{array}{ccccc} 0 & \longrightarrow & L(\tau) & \longrightarrow & R \\ & & \downarrow{\scriptstyle\alpha} & & \downarrow{\scriptstyle\beta} \\ & & \bar{M} & \longrightarrow & E_\tau(\bar{M}) = Q_\tau(M) \end{array}$$

commute. Define the map $\varphi\colon \mathrm{Hom}_R(L(\tau),\bar{M}) \to E_\tau(\bar{M}) = Q_\tau(M)$ by $\varphi\colon \alpha \mapsto 1\beta$. By the uniqueness of β, φ is a well-defined R-homomorphism. Also, φ is clearly monic. Moreover, if $x \in E_\tau(\bar{M})$, then x determines an R-homomorphism $\beta\colon R \to E_\tau(\bar{M})$ defined by $\beta\colon r \mapsto rx$. Let $\nu\colon E_\tau(\bar{M}) \to E_\tau(\bar{M})/\bar{M}$ be the canonical R-epimorphism. Then $E_\tau(\bar{M})/\bar{M}$ is τ-torsion and so $R\beta\nu \in \mathcal{J}_\tau$. Therefore, $\ker(\beta\nu) \in \mathcal{L}_\tau$ and so contains $L(\tau)$. Thus, if α is the restriction of β to $L(\tau)$, $\mathrm{im}(\alpha) \subseteq \bar{M}$. Since $1\beta = \varphi(\alpha)$, we see that φ is epic and so is an isomorphism. □

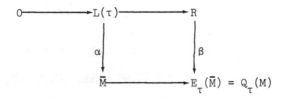

(22.7) PROPOSITION: The following conditions are equivalent for $\tau \in$ R-tors:

(1) τ is jansian.

(2) $Q_\tau(\_)$ commutes with direct products.

PROOF: (1) ⇒ (2): If $\{M_i\}$ is a family of left R-modules, then by Proposition 22.6, $Q_\tau(\Pi M_i) \cong$ $\mathrm{Hom}_R(L(\tau),[\Pi M_i]/T_\tau(\Pi M_i)) \cong \mathrm{Hom}_R(L(\tau),\Pi[M_i/T_\tau(M_i)]) \cong$ $\Pi\mathrm{Hom}_R(L(\tau),M_i/T_\tau(M_i)) \cong \Pi Q_\tau(M_i)$.

(2) ⇒ (1): Let $\{M_i\}$ be a family of τ-torsion left R-modules. Then $Q_\tau(\Pi M_i) \cong \Pi Q_\tau(M_i) = 0$ and so ΠM_i is τ-torsion. Therefore (1) follows by Proposition 22.2. □

(22.8) PROPOSITION: Let $\tau \in$ R-jans. Then there exists a $\tau^* \in$ R-tors satisfying the condition that $\mathcal{J}_{\tau^*} = \{M \in \text{R-mod} \mid m \in L(\tau)m$ for all $m \in M\}$.

PROOF: Let $\mathcal{A} = \{M \in \text{R-mod} \mid m \in L(\tau)m$ for all $m \in M\}$. Then $\mathcal{A}$ is clearly closed under taking submodules and homomorphic images. If $0 \to M' \xrightarrow{\alpha} M \xrightarrow{\beta} M'' \to 0$ is exact with $M', M'' \in \mathcal{A}$ and if $m \in M \setminus M'\alpha$, then there exists an $a \in L(\tau)$ for which $m\beta = am\beta$. Therefore, $am - m \in \ker(\beta) = \mathrm{im}(\alpha)$ and so there exists a $b \in L(\tau)$ for which $am - m = b(am - m)$. Hence $m = am + bm - abm \in L(\tau)m$ and so $M \in \mathcal{A}$.

Finally, let $M = \oplus M_i$ where each $M_i \in \mathcal{A}$. Let $m = \langle m_i \rangle \in M$ with nonzero entries $m_{i_1},\ldots,m_{i_n}$. Then for each positive integer j, with $1 \leq j \leq n$, there exists an $a_j \in L(\tau)$ with $a_j m_{i_j} = m_{i_j}$. We claim that there exists an $a \in L(\tau)$ with $am_{i_j} = m_{i_j}$ for $1 \leq j \leq n$. This will be proven by induction on n.

For $n = 1$, the result is trivial. Assume, therefore, that for any $x_{i_j} \in M_{i_j}$ $(1 \leq j \leq n-1)$ we can find a $b \in L(\tau)$ with $bx_{i_j} = x_{i_j}$ for $1 \leq j \leq n-1$. In particular, we can then find an $a' \in L(\tau)$ such that $a'(m_{i_j} - a_n m_{i_j}) = m_{i_j} - a_n m_{i_j}$ for $1 \leq j \leq n-1$. Set $a = a' + a_n - a'a_n \in L(\tau)$. A straightforward calculation shows that this is in fact the a that we seek. □

The theorem now follows by Proposition 1.6. □

Note that if $M \in \mathcal{J}_{\tau^*}$ then $T_\tau(M) \in \mathcal{J}_{\tau^*}$ and so $L(\tau)T_\tau(M) = T_\tau(M)$. On the other hand, $T_\tau(M) \in \mathcal{J}_\tau$ and so $L(\tau)T_\tau(M) = 0$. Therefore, $T_\tau(M) = 0$ and so we conclude that $\mathcal{J}_{\tau^*} \subseteq \mathcal{F}_\tau$. In particular, τ and τ^* are disjoint.

If $\tau \in R$-jans, then the torsion theory τ^* is called the <u>left associate</u> of τ.

(22.9) PROPOSITION: <u>Let</u> $\tau \in R$-jans. <u>Then</u> $\mathcal{L}_{\tau^*} = \{{}_R I \subseteq R \mid (I:r) + L(\tau) = R$ <u>for all</u> $r \in R\}$.

PROOF: If $I \in \mathcal{L}_{\tau^*}$ and $r \in R$, then $(I:r) \in \mathcal{L}_{\tau^*}$ and so $(I:r) + L(\tau) = R$ by Proposition 8.8(2). Conversely, if $(I:r) + L(\tau) = R$ for all $r \in R$, then given $r \in R$, we have $1 = b + a$ for $b \in (I:r)$ and $a \in L(\tau)$. Therefore, $r + I = b(r + I) + a(r + I) = a(r + I)$ and so $r + I \in L(\tau)(r + I)$. This shows that $R/I \in \mathcal{J}_{\tau*}$ and so $I \in \mathcal{L}_{\tau*}$. □

(22.10) PROPOSITION: <u>The following conditions are equivalent for</u> $\tau \in$ R-jans:

(1) $R/L(\tau)$ <u>is flat as a right</u> R-<u>module</u>.

(2) <u>There exists a</u> $\tau' \in$ R-tors <u>with</u> $\mathcal{J}_\tau = \mathcal{F}_{\tau'}$.

(3) $\mathcal{J}_\tau = \mathcal{F}_{\tau*}$.

(4) <u>For any</u> $a_1, \ldots, a_n \in L(\tau)$, <u>we have</u> $\bigcap_i (0:a_i) + L(\tau) = R$.

(5) <u>For any</u> $a \in L(\tau)$, <u>we have</u> $(0:a) + L(\tau) = R$.

(6) τ <u>is stable</u>.

(7) $\mathcal{J}_{\tau*} = \{M \in \text{R-mod} \mid M = L(\tau)M\}$.

PROOF: (1) ⇒ (2): By Proposition 1.6, $\mathcal{J}_\tau$ is closed under taking submodules and homomorphic images. By Proposition 22.2, it is closed under taking direct products. Since $R/L(\tau)$ is flat as a right R-module, $I \cap L(\tau) = L(\tau)I$ for every left ideal I of R [89, p. 133]. Let M be a τ-torsion left R-module and let $0 \neq x \in E(M)$. Assume that $L(\tau)x \neq 0$. Since M is large in $E(M)$, this means that $M \cap L(\tau)x \neq 0$ and so there exists an $a \in L(\tau) \cap (M:x)$ with $ax \neq 0$. But if $a \in L(\tau) \cap (M:x)$, then $a \in L(\tau)(M:x)$ and so $ax \in L(\tau)(M:x)x \subseteq L(\tau)M = 0$. Therefore, $L(\tau)x = 0$ for all $x \in E(M)$ and so $E(M)$ is τ-torsion. The result then follows by Proposition 1.4.

(2) ⇒ (3): By Proposition 1.3, we must have $\mathcal{J}_{\tau'} = \{M \in R\text{-mod} \mid L(\tau)M = M\}$ and so in particular $\mathcal{J}_{\tau*} \subseteq \mathcal{J}_{\tau'}$. Conversely, if $M \in \mathcal{J}_{\tau'}$, then for each $m \in M$, we have $Rm \in \mathcal{J}_{\tau'}$ and so $m \in Rm = L(\tau)Rm = L(\tau)m$. Thus $M \in \mathcal{J}_{\tau*}$ and so $\tau' = \tau*$.

(3) ⇒ (4): Since $L(\tau)$ is idempotent, $L(\tau) \in \mathcal{J}_{\tau*}$ by (3) and so $(0:a) \in \mathcal{L}_{\tau*}$ for every $a \in L(\tau)$. Thus $\cap(0:a_i) \in \mathcal{L}_{\tau*}$ and so, by Proposition 22.9, $\cap(0:a_i) + L(\tau) = R$.

(4) ⇒ (5): The proof is trivial.

(5) ⇒ (1): Let $a \in L(\tau)$. Then by (5) we can write $1 = a' + b$ for some $a' \in L(\tau)$ and $b \in (0:a)$. Then $a = a'a + ba = a'a \in L(\tau)a$. By [27], this suffices to prove that $R/L(\tau)$ is a flat right R-module.

(3) ⇒ (6): The proof is trivial.

(6) ⇒ (7): Let $\mathcal{A} = \{M \in R\text{-mod} \mid M = L(\tau)M\}$. Then $\mathcal{J}_{\tau*} \subseteq \mathcal{A}$ with equality if and only if $\mathcal{A}$ is closed under taking submodules. Let $M \in \mathcal{A}$. Let M' be a submodule of M and set $N = M'/L(\tau)M' \in \mathcal{J}_{\tau}$. Since τ is stable, $E(N)$ is τ-torsion and we have the exact sequence $\mathrm{Hom}_R(M,E(N)) \to \mathrm{Hom}_R(M',E(N)) \to 0$. But $\mathrm{Hom}_R(M,E(N)) = 0$ since, for every R-homomorphism $\alpha: M \to E(N)$, we have $M\alpha \in \mathcal{A}$

and so $M\alpha = L(\tau)M\alpha$, while on the other hand, $M\alpha \subseteq E(N)$ implies that $M\alpha$ is τ-torsion and so $L(\tau)M\alpha = 0$. Therefore, $\text{Hom}_R(M',E(N)) = 0$. In particular, this means that $M'/L(\tau)M' = 0$ and so $M' \in \mathcal{A}$.

(7) ⇒ (3): The proof is the same as for (2) ⇒ (3). □

(22.11) COROLLARY: _If_ $\tau \in R\text{-jans}$ _is stable, then_ $L(\tau) = T_{\tau*}(R)$.

PROOF: By Proposition 22.10(7), $L(\tau) \in \mathcal{J}_{\tau*}$ and so $L(\tau) \subseteq T_{\tau*}(R)$. Conversely, if $a \in T_{\tau*}(R)$, then $a \in L(\tau)a \subseteq L(\tau)$ and so we have equality. □

(22.12) PROPOSITION: _The following conditions are equivalent for_ $\tau \in R\text{-jans}$:

(1) $\mathcal{F}_\tau = \mathcal{J}_{\tau*}$.

(2) $M = L(\tau)M$ _for every_ $M \in \mathcal{F}_\tau$.

(3) $R = T_\tau(R) + L(\tau)$.

(4) $M = T_\tau(M) + L(\tau)M$ _for every left_ R-_module_ M.

(5) $R/L(\tau)$ _is projective as a left_ R-_module_.

(6) $\mathcal{F}_\tau$ _is closed under taking homomorphic images_.

PROOF: (1) ⇒ (2): If $m \in M \in \mathcal{F}_\tau$, then $Rm = L(\tau)m$ by (1) and so $M = L(\tau)M$.

(2) ⇒ (3): By (2), $R/T_\tau(R) = L(\tau)[R/T_\tau(R)] \cong L(\tau)/[L(\tau) \cap T_\tau(R)] \cong [L(\tau) + T_\tau(R)]/T_\tau(R)$ and so $R = T_\tau(R) + L(\tau)$.

(3) $\Leftrightarrow$ (4): Assume (3). Then for every left R-module M, we have $M = [T_\tau(R) + L(\tau)]M \subseteq T_\tau(R)M + L(\tau)M \subseteq T_\tau(M) + L(\tau)M \subseteq M$, proving (4). The converse is trivial.

(3) $\Rightarrow$ (5): By (3), we can write $1 = b + a$, where $b \in T_\tau(R)$ and $a \in L(\tau)$. If $a' \in L(\tau)$, then $a' = a'\cdot 1 = a'b + a'a = a'a$ since $a'b' = 0$ for all $b' \in T_\tau(R)$. Therefore, by [27, Proposition 2.2], $R/L(\tau)$ is a flat left R-module. Furthermore, the above also shows that $L(\tau) = Ra$ and so $R/L(\tau)$ is finitely presented, which, together with flatness, implies that $R/L(\tau)$ is projective as a left R-module.

(5) $\Rightarrow$ (6): Let M be a τ-torsion-free left R-module and let $\alpha: M \to N$ be an R-epimorphism. Since $R/L(\tau)$ is projective as a left R-module, we have an epimorphism $\mathrm{Hom}_R(R/L(\tau),M) \to \mathrm{Hom}_R(R/L(\tau),N)$. But $R/L(\tau)$ is τ-torsion and so $\mathrm{Hom}_R(R/L(\tau),M) = 0$. Therefore, $\mathrm{Hom}_R(R/L(\tau),N) = 0$, which suffices to prove that N is τ-torsion-free.

(6) $\Rightarrow$ (1): We have already noted that $\mathcal{T}_{\tau*} \subseteq \mathcal{F}_\tau$, so all we have to show is the reverse containment. If $m \in M \in \mathcal{F}_\tau$, then $Rm \in \mathcal{F}_\tau$, and so by (6) $Rm/L(\tau)m \in \mathcal{F}_\tau$. But $L(\tau)[Rm/L(\tau)m] = 0$ and so $Rm/L(\tau)m \in \mathcal{T}_\tau$ whence we must have $Rm = L(\tau)m$. This implies that $Rm \in \mathcal{T}_{\tau*}$. Since this is true for any $m \in M$, we therefore have $M \in \mathcal{T}_{\tau*}$. □

We should also recall that torsion theories satisfying

condition (6) of Proposition 22.12 were characterized in Proposition 5.5.

(22.13) PROPOSITION: <u>The following conditions are equivalent for</u> $\tau \in$ R-jans:

(1) $\mathcal{T}_\tau = \mathcal{F}_{\tau*}$ <u>and</u> $\mathcal{F}_\tau = \mathcal{T}_{\tau*}$.

(2) $R = T_\tau(R) \times T_{\tau*}(R)$. <u>(as rings)</u>

PROOF: (1) $\Rightarrow$ (2): The ideal $T_\tau(R) \cap T_{\tau*}(R)$ is a submodule of both $T_\tau(R)$ and $T_{\tau*}(R)$ and so, by (1), belongs to $\mathcal{T}_\tau \cap \mathcal{F}_\tau = \{0\}$. The module $R/[T_\tau(R) + T_{\tau*}(R)]$ is a homomorphic image of both $R/T_\tau(R)$ and $R/T_{\tau*}(R)$ and so, by (1), belongs to $\mathcal{T}_{\tau*} \cap \mathcal{F}_{\tau*} = \{0\}$. Therefore, as left R-modules, we have $R = T_\tau(R) \oplus T_{\tau*}(R)$.

Write $1 = e_\tau + e_{\tau*}$ where $e_\tau \in T_\tau(R)$ and $e_{\tau*} \in T_{\tau*}(R)$. Then $e_\tau e_{\tau*}$ and $e_{\tau*}e_\tau$ belong to $T_\tau(R) \cap T_{\tau*}(R) = 0$. Indeed, for any $a \in T_\tau(R)$, we have $a = ae_\tau + ae_{\tau*} = ae_\tau$ since $ae_{\tau*} \in T_\tau(R) \cap T_{\tau*}(R) = 0$ so $T_\tau(R) = Re_\tau$. In particular we see that e_τ is idempotent. Similarly, $T_{\tau*}(R) = Re_{\tau*}$ and $e_{\tau*}$ is idempotent. Switching sides, a similar argument yields $T_\tau(R) = e_\tau(R)$ and $T_{\tau*}(R) = e_{\tau*}R$. Therefore, for any $a \in R$, we have $ae_\tau = e_\tau ae_\tau = e_\tau a$ and so e_τ is central. Likewise, $e_{\tau*}$ is central.

(2) $\Rightarrow$ (1): The decomposition $R = T_\tau(R) \times T_{\tau*}(R)$ yields a decomposition $1 = e_1 + e_2$ where e_1 and e_2 are central

idempotents of R, $e_1e_2 = e_2e_1 = 0$, and $e_1 \in T_\tau(R)$, $e_2 \in T_{\tau*}(R)$. We claim that $\mathcal{J}_\tau = \{M \in R\text{-mod} \mid e_2M = 0\}$. Indeed, for any left R-module M and any $m \in M$, we have that $e_2m = 0$ implies that $m = e_1m$ and so $m \in T_\tau(R)m \subseteq T_\tau(M)$. Conversely, if $m \in M \in \mathcal{J}_\tau$, then $e_2 \in T_{\tau*}(M)$ implies that $e_2 \in L(\tau)e_2 = e_2L(\tau)$ by the centrality of e_2 and so $e_2m \in e_2L(\tau)m = 0$.

A parallel argument shows that $\mathcal{J}_{\tau*} = \{M \in R\text{-mod} \mid e_1M = 0\}$. Moreover, it is easily seen that $\{M \in R\text{-mod} \mid e_2M = 0\}$ satisfies the conditions of Proposition 1.4(2) and so there exists a $\tau' \in R\text{-tors}$ with $\mathcal{J}_\tau = \mathcal{F}_{\tau'}$. Then $\mathcal{J}_\tau = \mathcal{F}_{\tau*}$ by Proposition 22.10. Also, if $x \in M \in \mathcal{F}_\tau$, then $x = e_1x + e_2x$. Since $e_2e_1x = 0$, then $e_1x \in T_\tau(M) = 0$. Hence $e_1M = 0$. Conversely, if $e_1M = 0$ and if $m \in T_\tau(M)$, then $m = e_1m + e_2m = 0$. Thus $\mathcal{F}_\tau = \{M \in R\text{-mod} \mid e_1M = 0\} = \mathcal{J}_{\tau*}$, proving (1). □

A jansian torsion theory satisfying the conditions of Proposition 22.13 is said to be *centrally splitting.*

(22.14) PROPOSITION: *The following conditions are equivalent for* $\tau \in R\text{-jans}$:

(1) τ *is centrally splitting*.

(2) *For any left* R-*module* M, $T_{\tau*}(M) \in \mathcal{F}_\tau$ *and* $M/T_\tau(M) \in \mathcal{J}_{\tau*}$.

(3) *For any left* R-*module* M, $M = T_\tau(M) \oplus T_{\tau*}(M)$.

(4) τ *is stable and* $T_{\tau*}(R)$ *is a direct summand of* ${}_RR$.

(5) τ *is stable and* $R/L(\tau)$ *has a projective cover in* R-mod.

(6) τ *is stable and* $T_\tau(\_)$ *is exact*.

PROOF: (1) ⇒ (2): This follows from Proposition 22.13(1).

(2) ⇒ (3): Let M be a left R-module. Then $T_\tau(M) \cap T_{\tau*}(M) = 0$ by (2) and $M/[T_\tau(M) + T_{\tau*}(M)] = 0$ by (2) since it is a homomorphic image of $M/T_\tau(M) \in \mathcal{T}_{\tau*}$ and $M/T_{\tau*}(M) \in \mathcal{F}_{\tau*}$. Therefore, $M = T_\tau(M) \oplus T_{\tau*}(M)$.

(3) ⇒ (4): Let M be a τ-torsion left R-module. By (3), $E(M) = T_\tau(E(M)) \oplus T_{\tau*}(E(M))$. Since $M \subseteq T_\tau(E(M))$ is large in E(M), we must therefore have $E(M) = T_\tau(E(M)) \in \mathcal{T}_\tau$. Thus τ is stable. Also, by (3), $T_{\tau*}(R)$ is a direct summand of ${}_RR$.

(4) ⇒ (5): This follows trivially, since by Proposition 22.10 and Corollary 22.11, $L(\tau) = T_{\tau*}(R)$.

(5) ⇒ (1): Let $\mu: P \to R/L(\tau)$ be a projective cover and let $\nu: R \to R/L(\tau)$ be the canonical surjection. Then by the projectivity of R, there exists an R-homomorphism $\alpha: R \to P$ with $\alpha\mu = \nu$. Since ν is an R-epimorphism, this means that $R\alpha + \ker(\mu) = P$ and so, by the smallness of

$\ker(\mu)$ in P, we have $R\alpha = P$. Since P is projective, α splits and so $P \cong Re$ for some idempotent e in R. Therefore, the restriction ν' of ν to Re is also a projective cover of $R/L(\tau)$. In particular, this implies that $Re + L(\tau) = R$ and that $\ker(\nu') = Re \cap L(\tau) = L(\tau)e$ is a small submodule of Re.

Let $a \in L(\tau)$ and let $\beta\colon Re \to Rea$ be the R-homomorphism defined by $x \mapsto xa$. Then by Proposition 22.10(7), $ea \in L(\tau) \in \mathfrak{J}_{\tau*}$ and so $Rea = L(\tau)ea = [L(\tau)e]\beta$. Therefore, $Re = L(\tau)e + \ker(\beta)$. By the smallness of $L(\tau)e$ in Re, this means that $Re = \ker(\beta)$ and so $Rea = 0$ for every $a \in L(\tau)$. Hence $ReL(\tau) = 0$.

Moreover, $ReR = Re(Re + L(\tau)) = ReRe + ReL(\tau) = ReRe \subseteq Re$ and so Re is a two-sided ideal of R. Since $R/L(\tau)$ is flat as a right R-module, this means that $L(\tau)e = L(\tau) \cap Re = ReL(\tau) = 0$ [89, p. 133]. Hence $e \in T_\tau(R)$ and so $R = T_\tau(R) + L(\tau)$, which proves (1) by Proposition 22.12.

(1) ⇒ (6): This follows from Propositions 22.12 and 5.5, and from (1) ⇔ (4).

(6) ⇒ (1): This follows from Propositions 22.12 and 5.5, and from (1) ⇔ (3). □

(22.15) PROPOSITION: Let $\tau \in R$-tors and let $\tau' = \xi(T_\tau(R))$. Then the following conditions are equivalent:

(1) τ' is centrally splitting.

(2) $T_\tau(R)$ is generated by a central idempotent.

PROOF: (1) $\Rightarrow$ (2): Clearly $\tau' \leq \tau$ and so $T_{\tau'}(R) \subseteq T_\tau(R)$. But since $T_\tau(R)$ is τ'-torsion, we in fact have equality. By Proposition 22.13, (1) implies that $T_{\tau'}(R)$ is then generated by a central idempotent, proving (2).

(2) $\Rightarrow$ (1): Let $R = T_\tau(R) \oplus I$. By (2), I is an idempotent two-sided ideal of R and so I generates $\tau'' \in R\text{-jans}$ defined by $\mathcal{J}_{\tau''} = \{M \in R\text{-mod} \mid IM = 0\}$. Moreover, $I = T_{\tau''}(R)$ and so, by Proposition 22.13, τ'' is centrally splitting.

Since $IT_\tau(R) = 0$, $T_\tau(R)$ is τ''-torsion and so $\tau' \leq \tau''$. On the other hand, if M is a τ''-torsion left R-module, then $IM = 0$ and so $T_\tau(R)M = M$. Therefore, M is a homomorphic image of a direct sum of copies of $T_\tau(R)$ and so M is τ'-torsion. Thus $\tau'' \leq \tau'$ and so we have equality, which proves (1). □

(22.16) PROPOSITION: Let $R = \times_{i=1}^{n} R_i$ be a direct product of finitely-many rings. Then the following conditions are equivalent:

(1) Every $\tau \in R\text{-tors}$ is centrally splitting.

(2) Every $\sigma \in R_i\text{-tors}$ is centrally splitting, with $1 \leq i \leq n$.

PROOF: (1) ⇒ (2): Pick i, $1 \leq i \leq n$, and let $\lambda: R_i \to R$ be the canonical ring embedding and $\pi: R \to R_i$ be the canonical ring surjection. Let $\sigma \in R_i$-tors and set $\tau = \lambda_{\#}(\sigma) \in R$-tors. Then $\sigma = \pi_{\#}(\tau)$. By (1), τ is centrally splitting and so there exists a central idempotent $e \in R$ with $\mathcal{J}_\tau = \{M \in R\text{-mod} \mid eM = M\}$ and $\mathcal{F}_\tau = \{M \in R\text{-mod} \mid eM = 0\}$. Let $e' = \pi(e) \in R_i$. Then e' is a central idempotent of R_i and $\mathcal{J}_\sigma = \{N \in R_i\text{-mod} \mid {}_RN \in \mathcal{J}_\tau\} = \{N \in R_i\text{-mod} \mid e'N = N\}$. Similarly, $\mathcal{F}_\sigma = \{N \in R_i\text{-mod} \mid e'N = 0\}$. This suffices to show that σ is centrally splitting.

(2) ⇒ (1): For $1 \leq i \leq n$, let $\pi_i: R \to R_i$ be the canonical ring surjections. Let $\tau \in R$-tors and for each i let $\sigma_i = (\pi_i)_{\#}(\tau) \in R_i$-tors. Then each σ_i is centrally splitting and so, by Proposition 22.13, $R_i = T_{\sigma_i}(R_i) \times T_{\sigma_i^*}(R_i)$. Therefore, $R = [\times T_{\sigma_i}(R_i)] \times [\times T_{\sigma_i^*}(R_i)] = T_\tau(R) \times T_{\tau^*}(R)$, and so τ is centrally splitting. □

In Chapter VII, Example 31 we present an example of a jansian torsion theory that is not perfect. It is, therefore, of interest to characterize the perfect jansian theories.

(22.17) PROPOSITION: <u>The following conditions are equivalent for</u> $\tau \in R$-tors:

(1) τ <u>is perfect and jansian</u>.

(2) τ <u>is perfect and</u> $(R_\tau)_R$ <u>is finitely generated and projective</u>.

(3) τ <u>is perfect and</u> $Q_\tau(\_)$ <u>commutes with direct products</u>.

(4) (i) τ <u>is jansian</u>;

(ii) ${}_RL(\tau)$ <u>is finitely generated; and</u>

(iii) ${}_RL(\tau)$ <u>is</u> τ-<u>projective</u>.

PROOF: (1) $\Rightarrow$ (4): Since $\tau \in$ R-perf, $\mathcal{L}_\tau$ has a cofinal subset of finitely generated left ideals, by Proposition 17.1. Therefore, since $L(\tau)$ is a minimal element of $\mathcal{L}_\tau$, it must be finitely generated as a left ideal of R. Furthermore, if $\alpha: M \to N$ is an R-epimorphism, then by Proposition 17.1, $Q_\tau(\alpha): Q_\tau(M) \to Q_\tau(N)$ is an R-epimorphism. If M, N are τ-torsion-free, then by Proposition 22.6, $Q_\tau(M) \cong \mathrm{Hom}_R(L(\tau),M)$ and $Q_\tau(N) \cong \mathrm{Hom}_R(L(\tau),N)$ and so (4iii) follows directly.

(4) $\Rightarrow$ (3): By (4ii), τ is clearly of finite type. By Proposition 22.7, $Q_\tau(\_)$ commutes with direct products. It therefore remains to show, by Proposition 17.1, that τ is exact.

Let $I \in \mathcal{L}_\tau$ and let $\beta: M \to N$ be an R-epimorphism with $M \in \mathcal{E}_\tau$ and $N \in \mathcal{F}_\tau$. If $\alpha: I \to N$ is an R-homomorphism, then by the τ-projectivity of $L(\tau)$ there exists an R-homomorphism

φ making the diagram

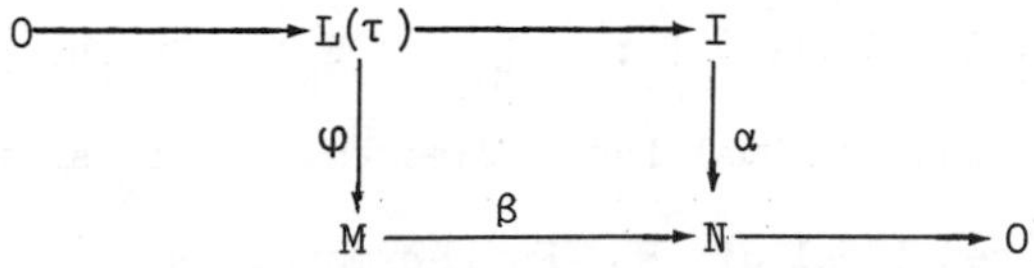

commute. Since $L(\tau)$ is τ-dense in I, then by the τ-injectivity of M there exists an R-homomorphism $\psi: I \to M$ making the diagram

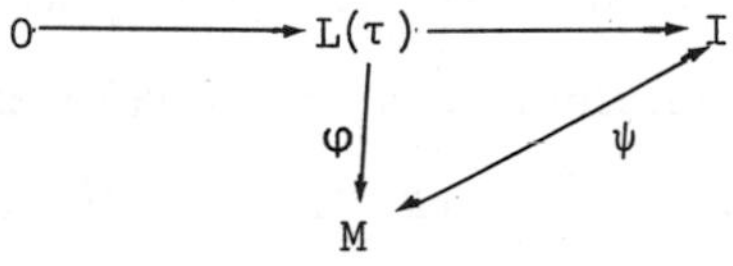

commute. Finally, the R-homomorphism $\alpha - \psi\beta$ induces an R-homomorphism $I/L(\tau) \to N$ which must be the zero map since N is τ-torsion-free. Therefore $\alpha = \psi\beta$, proving that I is τ-projective. By Proposition 16.3, this proves that τ is exact.

(3) ⇒ (2): Since τ is perfect, $(R_\tau)_R$ is flat. Moreover, $Q_\tau(\_)$ is naturally equivalent to $R_\tau \otimes_R \_$ and commutes with direct products. Therefore, by [31, Lemmas 1.1 and 1.2], $(R_\tau)_R$ is finitely generated and in fact is finitely presented. But finitely presented flat left R-modules are projective [27].

(2) ⇒ (1): Since τ is perfect, then by Proposition

17.1, a left R-module M is τ-torsion if and only if $0 = Q_\tau(M) \cong R_\tau \otimes_R M$. Let L be the trace ideal of $(R_\tau)_R$, i.e., $L = \Sigma\{\mathrm{im}(\alpha) \mid \alpha: (R_\tau)_R \to R_R\}$. By the projectivity of $(R_\tau)_R$ and the Dual Basis Theorem, L is an idempotent two-sided ideal of R. Furthermore, M is τ-torsion if and only if $LM = 0$. Therefore, $I \in \mathcal{L}_\tau$ if and only if $LR = L \subseteq I$. By Proposition 22.1, this proves that $\tau \in R\text{-jans}$ with $L = L(\tau)$. □

Let $\tau \in R\text{-tors}$ and let $\gamma: R \to R/T_\tau(R) = S$ be the canonical ring surjection. Set $\sigma = \gamma_\#(\tau) \in S\text{-tors}$. By Proposition 17.20, $\tau \in R\text{-perf}$ if and only if $\sigma \in S\text{-perf}$. Moreover, $\mathcal{L}_\sigma = \{[H + T_\tau(R)]/T_\tau(R) \mid H \in \mathcal{L}_\tau\}$ so that $\tau \in R\text{-jans}$ if and only if $\sigma \in S\text{-jans}$. Therefore, we can reduce the study of perfect jansian torsion theories to the study of faithful perfect jansian torsion theories. In this case, we can strengthen condition (4) of Proposition 22.17.

(22.18) PROPOSITION: <u>The following conditions are equivalent for a faithful</u> $\tau \in R\text{-tors}$:

(1) τ <u>is perfect and jansian</u>.

(2) τ <u>is jansian and</u> ${}_R L(\tau)$ <u>is finitely generated and projective</u>.

<u>Moreover, in this case,</u> R_τ <u>is isomorphic to the endomorphism ring of</u> $L(\tau)$.

PROOF: By Proposition 22.17, (2) implies (1). Conversely, assume (1). Then in particular, τ is exact and so, by Proposition 16.3, $L(\tau)$ is τ-projective. Let M be a free left R-module for which there exists an R-epimorphism $\alpha: M \to L(\tau)$. Since τ is faithful, M is τ-torsion-free and $L(\tau)$, being a submodule of R, is τ-torsion-free. Therefore, by τ-projectivity there exists an R-homomorphism θ making the diagram

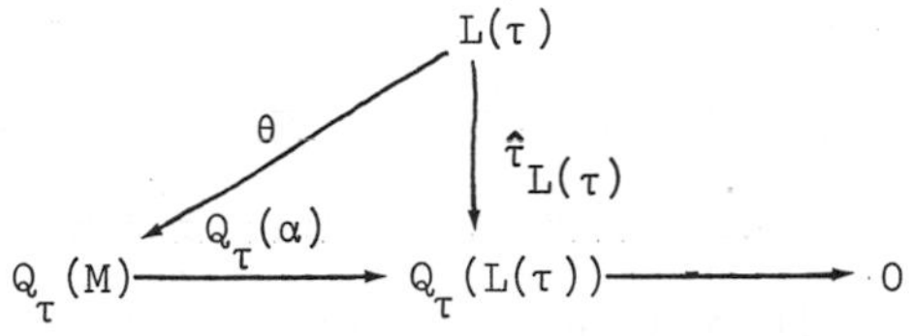

commute. Since $Q_\tau(M)$ is τ-injective, θ extends to an R-homomorphism $\psi: R \to Q_\tau(M)$. Let $q = 1\psi \in Q_\tau(M)$. Since M is τ-dense in $Q_\tau(M)$, there exists an $I \in \mathcal{L}_\tau$ with $Iq \subseteq M$. By the minimality of $L(\tau)$, we have $I \supseteq L(\tau)$ and so $L(\tau)q \subseteq M$, i.e., $\mathrm{im}(\theta) \subseteq M$. Therefore, $\theta\alpha$ is the identity map on $L(\tau)$ and so α splits, proving that $L(\tau)$ is projective.

As for the last statement, by the projectivity of $L(\tau)$ we have the exact sequence $0 \to \mathrm{Hom}_R(L(\tau),L(\tau)) \to \mathrm{Hom}_R(L(\tau),R) \to \mathrm{Hom}_R(L(\tau),R/L(\tau)) \to 0$. But, since $L(\tau)$ is idempotent, $\mathrm{Hom}_R(L(\tau),R/L(\tau)) = 0$.

Thus, by Proposition 22.6, $R_\tau \cong Q_\tau(R) = \mathrm{Hom}_R(L(\tau),R) \cong \mathrm{Hom}_R(L(\tau),L(\tau))$. □

(22.19) PROPOSITION: <u>Let</u> $\mathcal{A} \subseteq$ R-simp <u>and let</u> $\tau = \xi(\mathcal{A})$ <u>be a semisimple torsion theory. Suppose that for each</u> $M \in$ R-simp $\smallsetminus \mathcal{A}$ <u>there exists an idempotent</u> $e_M \in R$ <u>satisfying</u> $e_M M \neq 0$ and $e_M M' = 0$ <u>for all</u> $M \neq M' \in$ R-simp. <u>If</u> ξ(R-simp) <u>is jansian then</u> τ <u>is jansian</u>.

PROOF: Let $\{N_k \mid k \in \Omega\}$ be a family of τ-torsion left R-modules and let $N = \Pi N_k$. By hypothesis, ξ(R-simp) is jansian and so N is ξ(R-simp)-torsion. By Propositions 2.6 and 2.7, there then exists an ordinal j with $F_j(N) = N$, where $\langle F_i \rangle$ is the Loewy sequence on R-mod. Let $M \in$ R-simp $\smallsetminus \mathcal{A}$ and let $k \in \Omega$. Then for any nonlimit ordinal i, no submodule of $F_i(N_k)/F_{i-1}(N_k)$ is isomorphic to M. We claim, therefore, that $e_M N_k = 0$. Indeed, assume that $e_M N_k \neq 0$. Then there exists an $x \in N_k$ with $e_M x \neq 0$. Pick an ordinal i that is minimal with respect to $e_M x \in F_i(N_k)$. Then i is not a limit ordinal and $e_M x \notin F_{i-1}(N_k)$. If M is not isomorphic to a submodule of $F_i(N_k)/F_{i-1}(N_k)$, then, since $F_i(N_k)/F_{i-1}(N_k)$ is a direct sum of simple left R-modules not isomorphic to M, we have $e_M F_i(N_k) \subseteq F_{i-1}(N_k)$. But then $e_M^2 x = e_M x \in F_i(N_k) \smallsetminus F_{i-1}(N_k)$, which is a contradiction. Therefore, $e_M N_k = 0$.

Since the preceding is true for all $k \in \Omega$, we have $e_M N = 0$. We now claim that for every nonlimit ordinal i, no submodule of $F_i(N)/F_{i-1}(N)$ is isomorphic to M. Indeed, if $F_i(N)/F_{i-1}(N)$ has a submodule isomorphic to M, then $0 \neq e_M[F_i(N)/F_{i-1}(N)]$ and so $e_M F_i(N) \not\subseteq F_{i-1}(N)$, implying that $e_M N \neq 0$, which is a contradiction.

Since the above is true for each $M \in R\text{-simp} \setminus \mathcal{A}$, we see that for every nonlimit ordinal i, the module $F_i(N)/F_{i-1}(N)$ is isomorphic to a direct sum of copies of elements of $\mathcal{A}$. This suffices to prove that N is τ-torsion. By Proposition 22.2, this implies that $\tau \in R\text{-jans}$. □

(22.20) COROLLARY: _If_ R _is a semiperfect ring and if_ $\xi(R\text{-simp})$ _is jansian, then every semisimple_ $\tau \in R\text{-tors}$ _is jansian_.

PROOF: If R is semiperfect, then $R/J(R)$ is completely reducible and $R/J(R)$-simp is finite, say it is $\{M_1,\ldots,M_n\}$. Let $e_1',\ldots,e_n'$ be idempotent elements of $R/J(R)$ satisfying

$$e_i'M_j = \begin{cases} M_i & \text{if } i = j \\ 0 & \text{otherwise} \end{cases}$$

Let $e_1,\ldots,e_n$ be idempotents of R such that $e_i + J(R) = e_i'$ for each i. Then the M_i, considered as left R-modules, are also simple and

$$e_iM_j = \begin{cases} M_i & \text{if } i = j \\ 0 & \text{otherwise} \end{cases}$$

The result now follows from Proposition 22.19. □

(22.21) PROPOSITION: <u>Suppose that every semisimple</u> $\tau \in$ R-tors <u>is jansian. Then</u> $R/J(R)$ <u>is completely reducible</u>.

PROOF: Let $\tau_0 = \xi(\text{R-simp})$. For each $M \in$ R-simp, let $H_M = \cap\{_RI \subseteq R \mid R/I \cong M\}$. Then R/H_M can be embedded in a direct product of copies of M. Since $\xi(M)$ is jansian, then by Proposition 22.2, we see that $H_M \in \mathcal{L}_{\xi(M)}$.

Since the Jacobson radical of R is the intersection of all maximal left ideals of R, we have

$$(*) \qquad J(R) = \cap\{H_M \mid M \in \text{R-simp}\}.$$

For each $M \in$ R-simp, let $\bar{H}_M = \cap\{H_{M'} \mid M \neq M' \in \text{R-simp}\}$. Then $\bar{H}_M \in \mathcal{L}_{\vee\{\xi(M') \mid M \neq M' \in \text{R-simp}\}}$ which is contained in $\mathcal{L}_{\chi(M)}$ by Proposition 12.2. This shows that the representation (*) is irredundant, for if $J(R) = \bar{H}_M$ for some $M \in$ R-simp, then $R/J(R)$ would be $\chi(M)$-torsion and so M would be $\chi(M)$-torsion, which is a contradiction.

We have an embedding $R/J(R) \to \Pi\{R/I \mid I$ a maximal left ideal of $R\}$. Since each such I is τ_0-dense in R, and so τ_0 is jansian by hypothesis, we have $R/J(R)$ τ_0-torsion. Every submodule of $R/J(R)$, therefore, has a nonzero socle and so $K/J(R) = \mathrm{soc}(R/J(R))$ is a large left ideal of R. We want to show that in fact $K = R$. Assume that this is not so. Then there is a maximal left ideal I_0 of R containing K, where $R/I_0 \cong M_0$ for some $M_0 \in R$-simp. Since $K/J(R)$ is large in $R/J(R)$, then $I_0/J(R)$ is large in $R/J(R)$.

We claim that $K \subseteq H_{M_0}$. Indeed, let I_1 be a maximal left ideal of R with $R/I_1 \cong R/I_0$. To show that $K \subseteq I_1$ it suffices to show that $K/J(R) \subseteq I_1/J(R)$. To do this, it suffices to show that $I_1/J(R)$ is large in $R/J(R)$, since the socle of a ring is the intersection of its large left ideals. Assume that this not the case. Then there exists a nonzero left ideal I_1' of R properly containing $J(R)$ such that $I_1/J(R) \cap I_1'/J(R) = 0$. By the maximality of I_1, we must then have $R = I_1 + I_1'$ and so $R/J(R) = I_1/J(R) \oplus I_1'/J(R)$. Therefore, $R/I_1 \cong I_1'/J(R)$ which is a direct summand of $R/J(R)$ and so is a projective left $R/J(R)$-module. Since $R/I_0 \cong R/I_1$, this implies that R/I_0 is a projective $R/J(R)$-module and so $I_0/J(R)$ is a direct summand of $R/J(R)$, contradicting the fact that $I_0/J(R)$ is large in $R/J(R)$. Therefore, $I_1 \supseteq K$ and so $K \subseteq H_{M_0}$.

Since the representation (*) is irredundant, $\bar{H}_{M_0} \neq \bar{H}_{M_0} \cap H_{M_0} = J(R)$. But on the other hand, since $J(R) \subseteq K \subseteq H_{M_0}$, then $\bar{H}_{M_0} \cap K = J(R)$ and so $\bar{H}_{M_0}/J(R) \cap K/J(R) = 0$. Since $K/J(R)$ is large in $R/J(R)$, this means that $\bar{H}_{M_0} = J(R)$, which is a contradiction. Therefore, $K = R$ and so $R/J(R)$ is completely reducible. □

(22.22) PROPOSITION: <u>The following conditions are equivalent for a ring</u> R:

(1) R <u>is right perfect</u>.

(2) R <u>is left semiartinian and</u> R-tors = R-jans.

PROOF: (1) ⇒ (2): If R is right perfect, then it is left semiartinian [9]. By Proposition 12.6, every $\tau \in$ R-tors is, therefore, semisimple. In particular, $\xi(\text{R-simp}) = \chi$ and so is jansian. Then R-tors = R-jans by Corollary 22.20.

(2) ⇒ (1): By [9] and the fact that R is left semiartinian, it suffices to show that $R/J(R)$ is completely reducible, which follows by Propositions 12.6 and 22.21. □

<u>References for Section 22</u>

Alin and Armendariz [4]; Azumaya [8]; Bernhard [13, 15]; Cunningham [33]; Dlab [40]; Jans [79]; Kurata [87]; Miller [102]; Rutter [133]; Teply [154].

23. DIRECT DECOMPOSITIONS

Let $\tau \in$ R-tors and let $\{\tau_i\} \subseteq$ R-tors. We say that the τ_i <u>directly decompose</u> τ, and we write $\tau = \oplus \tau_i$, if and only if $T_\tau(M) = \oplus T_{\tau_i}(M)$ for every left R-module M. As an example of the occurrence of direct decompositions, we consider the following result.

(23.1) PROPOSITION: <u>The following conditions are equivalent for</u> $\tau \in$ R-tors:

(1) τ <u>is centrally splitting</u>.

(2) <u>There exists a</u> $\tau' \in$ R-tors <u>with</u> $\chi = \tau \oplus \tau'$.

PROOF: (1) $\Rightarrow$ (2): By Proposition 22.14, we can take $\tau' = \tau^*$.

(2) $\Rightarrow$ (1): Assume that $\chi = \tau \oplus \tau'$. Then for any τ-torsion-free left R-module M, we have $M = T_\chi(M) = T_\tau(M) \oplus T_{\tau'}(M) = T_{\tau'}(M)$ and so $\mathcal{F}_\tau \subseteq \mathcal{J}_{\tau'}$. Conversely, for any τ'-torsion left R-module M, we have $M = T_\tau(M) \oplus T_{\tau'}(M) = T_\tau(M) \oplus M$. Thus $T_\tau(M) = 0$ and so M is τ-torsion-free. Thus we see that $\mathcal{F}_\tau = \mathcal{J}_{\tau'}$. A similar argument shows that $\mathcal{J}_\tau = \mathcal{F}_{\tau'}$. By Proposition 22.10 and 22.12, we see that τ is centrally splitting. □

The following result shows that direct decompositions are preserved by ring surjections.

(23.2) PROPOSITION: Let I be a two-sided ideal of a ring R and let $\gamma\colon R \to R/I = S$ be the canonical ring homomorphism. If $\tau = \tau' \oplus \tau''$ in R-tors, then $\gamma_{\#}(\tau) = \gamma_{\#}(\tau') \oplus \gamma_{\#}(\tau'')$ in S-tors.

PROOF: Let N be a left S-module. Then $T_{\tau}({}_RN) = T_{\tau'}({}_RN) \oplus T_{\tau''}({}_RN)$ by hypothesis. By Proposition 9.8(1), we have $T_{\gamma_{\#}(\tau)}(N) = T_{\gamma_{\#}(\tau')}(N) \oplus T_{\gamma_{\#}(\tau'')}(N)$ and so $\gamma_{\#}(\tau) = \gamma_{\#}(\tau') \oplus \gamma_{\#}(\tau'')$. □

(23.3) PROPOSITION: The following conditions are equivalent for $\tau \in$ R-jans:

(1) $\tau = \tau' \oplus \tau''$ for some $\tau', \tau'' \in$ R-tors.

(2) There exists a centrally splitting torsion theory in $R/L(\tau)$-tors.

PROOF: Let $\gamma\colon R \to R/L(\tau) = S$ be the canonical ring surjection.

(1) ⇒ (2): By Proposition 23.2, $\gamma_{\#}(\tau) = \gamma_{\#}(\tau') \oplus \gamma_{\#}(\tau'')$. But every left S-module belongs to $\mathcal{J}_{\gamma_{\#}(\tau)}$ by the remark after Proposition 22.1, and so (2) follows from (1) by Proposition 23.1.

(2) ⇒ (1): Let $\sigma' \in$ S-tors be centrally splitting. By Proposition 23.1, there exists a $\sigma'' \in$ S-tors such that every left S-module N can be written as $T_{\sigma'}(N) \oplus T_{\sigma''}(N)$. In particular, σ'' is also centrally splitting. Let $H' = L(\sigma')$ and $H'' = L(\sigma'')$. Then $R/L(\tau) = H'/L(\tau) \oplus H''/L(\tau)$.

Furthermore, R/H' and R/H'' are flat as right R-modules by Proposition 22.10 and a left S-module is σ'-torsion [respectively, σ''-torsion] if and only if it is a left R/H'-module [respectively, R/H''-module].

Therefore, there exist torsion theories τ' and $\tau'' \in$ R-tors with $\mathcal{J}_{\tau'} = \{M \in R\text{-mod} \mid H'M = 0\}$ and $\mathcal{J}_{\tau''} = \{M \in R\text{-mod} \mid H''M = 0\}$. Now let M be a τ-torsion left R-module. Then $M = RM = (H' + H'')M \subseteq H'M + H''M$. Since $H'H'' + H''H' \subseteq H' \cap H'' = L(\tau)$ and $L(\tau)M = 0$, we have $H'M \in \mathcal{J}_{\tau''}$ and $H''M \in \mathcal{J}_{\tau'}$. Thus $M = T_{\tau'}(M) + T_{\tau''}(M)$. Moreover, this sum is direct, for if $M \in \mathcal{J}_{\tau'} \cap \mathcal{J}_{\tau''}$ then $H'M = 0 = H''M$ and so $0 = (H' + H'')M = RM = M$. Thus $\tau = \tau' \oplus \tau''$. □

(23.4) PROPOSITION: _Let_ $\tau, \tau' \in$ R-tors _be disjoint and let_ $\tau'' = \tau \vee \tau'$. _For each left_ R-_module_ M, _let_ $\nu_M: M \to M/T_\tau(M)$ _and_ $\nu'_M: M \to M/T_{\tau'}(M)$ _be the canonical surjections. Then the following conditions are equivalent_:

(1) $\tau'' = \tau \oplus \tau'$.

(2) _For any left_ R-_module_ M, $\{m \in M \mid Rm\nu_M \in \mathcal{J}_{\tau'}\} = T_\tau(M) \oplus T_{\tau'}(M) = \{m \in M \mid Rm\nu'_M \in \mathcal{J}_\tau\}$.

(3) _For any left_ R-_module_ M, $T_\tau(M) \oplus T_{\tau'}(M)$ _is_ τ''-_pure in_ M.

(4) $\{M \in R\text{-mod} \mid M\nu_M \in \mathcal{J}_{\tau'}\} = \mathcal{J}_{\tau''} =$
$\{M \in R\text{-mod} \mid M\nu'_M \in \mathcal{J}_{\tau}\}$.

PROOF: (1) ⇒ (2): Let M be a left R-module and let $N = \{m \in M \mid Rm\nu_M \in \mathcal{J}_{\tau'}\}$. Then clearly $N \supseteq T_\tau(M) \oplus T_{\tau'}(M)$. Conversely, by Proposition 8.10, N is τ''-torsion and so, by (1), $N \subseteq T_{\tau''}(M) = T_\tau(M) \oplus T_{\tau'}(M)$. This proves the first equality; the second one follows similarly.

(2) ⇒ (3): Let M be a left R-module and let $N = T_\tau(M) \oplus T_{\tau'}(M)$. For $m + N \in T_\tau(M/N)$, consider the short exact sequence $0 \to N\nu'_M \to [Rm + N]\nu'_M \to [Rm + N]/N \to 0$. Then both $N\nu'_M$ and $[Rm + N]/N$ are τ-torsion and so $[Rm + N]\nu'_M \in \mathcal{J}_\tau$. By (2), this says that $Rm + N \subseteq N$ and so $m \in N$. Therefore, N is τ-pure in M. Similarly, N is τ'-pure in M. Therefore, $M/N \in \mathcal{F}_\tau \cap \mathcal{F}_{\tau'} = \mathcal{F}_{\tau''}$ and so N is τ''-pure in M.

(3) ⇒ (1): By Proposition 2.1, $T_{\tau''}(M) = T_\tau(M) \oplus T_{\tau'}(M)$ for every left R-module M, proving (1).

(4) ⇔ (2): This follows from the equivalence of (1) and (2). □

(23.5) PROPOSITION: <u>Let</u> $U \subseteq R\text{-tors}$ <u>be separated and let</u> $U = \cup U_j$ <u>be a partition of</u> U. <u>Set</u> $\tau = \vee U$ <u>and for each</u> j <u>set</u> $\tau_j = \vee U_j$. <u>Then the following conditions are equivalent</u>:

(1) $\tau = \oplus\, \tau_j$.

(2) For each j, $\tau = \tau_j \oplus [\bigvee_{i=j} \tau_i]$.

PROOF: (1) $\Rightarrow$ (2): By (1), $\tau \leq \tau_j \oplus [\bigvee_{i \neq j} \tau_i]$. Since the reverse inequality is trivial, we have equality.

(2) $\Rightarrow$ (1): For any left R-module M, let $F(M) = \oplus T_{\tau_i}(M)$. By Proposition 8.12, $\vee U = \xi(\mathcal{A})$, where $\mathcal{A} = \{M \in R\text{-mod} \mid F(M) = M\}$. Let $M \in \mathcal{J}_\tau$ and pick $m \in M$ satisfying $m + F(M) \in F(M/F(M))$. If $m \notin F(M)$, then we can assume without loss of generality that there exists an index j with $m + F(M) \in T_{\tau_j}(M/F(M))$. Therefore, there exists an $I \in \mathcal{L}_{\tau_j}$ with $Im \subseteq F(M)$. By hypothesis, we can write $m = m_1 + m_2$ where $m_1 \in T_{\tau_j}(M)$ and $m_2 \in T_{\tau'}(M)$ for $\tau' = \bigvee_{i \neq j} \tau_i$. Replacing I by $I \cap (0:m_1)$ if necessary, we can assume that $I \subseteq (0:m_1)$ and so $Im = Im_2 \subseteq F(M)$. By Proposition 8.8, $I + (0:m_2) = R$. Consequently, $m_2 \in F(M)$. Since $m_1 \in F(M)$ we thus have $m \in F(M)$, which is a contradiction. Thus $F(M/F(M)) = 0$. By Proposition 2.5, we then have $F(\_) = T_\tau(\_)$, proving (1). □

Let $\tau \in R\text{-tors}$. We say that τ has an <u>atomic direct decomposition</u> if and only if $\tau = \oplus \tau_i$ where each τ_i is an atom of R-tors, i.e., where each τ_i is simple. Then it is clear that τ has an atomic direct decomposition if and only if $\tau = \oplus\{\xi(M) \mid M \in R\text{-simp} \cap \mathcal{J}_\tau\}$.

(23.6) PROPOSITION: <u>The following conditions are equivalent</u>:

(1) ξ(R-simp) has an atomic direct decomposition.

(2) For each $M \in$ R-simp, $\xi(\text{R-simp}) = \xi(M) \oplus \xi(\text{R-simp} \setminus \{M\})$.

(3) If $Rx \in \mathcal{J}_{\xi(\text{R-simp})}$ has a simple homomorphic image M, then M is isomorphic to a submodule of Rx.

(4) If $M \in$ R-simp and $N \in \mathcal{J}_{\xi(\text{R-simp} \setminus \{M\})}$, then every exact sequence of the form $0 \to N \to N' \to M \to 0$ splits.

(5) Every $I \in \mathcal{L}_{\xi(\text{R-simp})}$ can be written as $\bigcap_{i=1}^{n} I_i$ where

(i) $I_i \in \mathcal{L}_{\xi(M_i)}$ for $M_i \in$ R-simp $(1 \leq i \leq n)$;

(ii) $i \neq j$ implies that $M_i \neq M_j$; and

(iii) $\bigcap_{j \neq i} I_j \not\subseteq I_i$ $(1 \leq i \leq n)$.

Moreover, this decomposition is unique in the sense that if $I = \bigcap_{i=1}^{m} I'$ where $I'_i \in \mathcal{L}_{\xi(M'_i)}$ for $M'_i \in$ R-simp, and if conditions (ii) and (iii) are satisfied for the M'_i, then $n = m$ and $\{M_1,\ldots,M_n\} = \{M'_1,\ldots,M'_n\}$.

(6) $\mathcal{L}_{\xi(\text{R-simp})} = \{{}_R I \subseteq R \mid$ there exist $M_1,\ldots,M_n \in$ R-simp and $I_i \in \mathcal{L}_{\xi(M_i)}$, $1 \leq i \leq n$, such that $I \supseteq \cap I_i\}$.

PROOF: (1) ⇔ (2): This follows from Proposition 23.5.

(1) ⇒ (3): Let Rx be a cyclic ξ(R-simp)-torsion left R-module and let $\alpha: Rx \to M$ be an R-epimorphism, where M is simple. Without loss of generality, we can assume that $M \in$ R-simp. By (1), $Rx = \oplus\{T_{\xi(M')}(Rx) \mid M' \in \text{R-simp}\}$ and so there is some $M' \in$ R-simp such that the restriction α' of α to $T_{\xi(M')}(Rx)$ is nonzero. Then $M = \text{im}(\alpha')$ and so M is $\xi(M')$-torsion, which implies by simplicity of M that $M = M'$. But $T_{\xi(M')}(Rx) \neq 0$ implies that Rx has a submodule isomorphic to M.

(3) ⇒ (4): Consider an exact sequence $0 \to N \to N' \overset{\alpha}{\to} M \to 0$ with $M \in$ R-simp and $N \in \mathcal{J}_{\xi(\text{R-simp} \setminus \{M\})}$. Select $x \in N' \setminus \ker(\alpha)$. Then $Rx\alpha = M$ by simplicity of M and so, by (3), there is a submodule M' of Rx isomorphic to M. Moreover, by the choice of N, we have $N \cap M' = 0$. Therefore the sequence splits.

(4) ⇒ (1): We need to show that for every left R-module N, $T_{\xi(\text{R-simp})}(N) = \oplus\{T_{\xi(M)}(N) \mid M \in \text{R-simp}\}$. We will prove this by transfinite induction on the Loewy length $k(N)$ of N. If $k(N) = 1$, the result is immediate.

Assume that $k(N) = i$, where i is not a limit ordinal, and that $T_{\xi(\text{R-simp})}(N') = \oplus\{T_{\xi(M)}(N') \mid M \in \text{R-simp}\}$

for all $N' \in R\text{-mod}$, with $k(N') \leq i-1$. Let $<F_i>$ be the Loewy sequence on R-mod. Then $k(F_{i-1}(N)) = i-1$ and so $F_{i-1}(N) = \oplus\{T_{\xi(M)}(F_{i-1}(N)) \mid M \in R\text{-simp}\}$. Since $N/F_{i-1}(N)$ is the sum of simple modules, it suffices to show that for each submodule N'' of N for which $N''/F_{i-1}(N)$ is simple, $N'' = \oplus\{T_{\xi(M)}(N'') \mid M \in R\text{-simp}\}$. But this follows from the fact that if $M''/F_{i-1}(N) \cong M' \in R\text{-simp}$, then the exact sequence $0 \to F_{i-1}(N) \to N'' \to M' \to 0$ splits by (4).

Finally, assume that $k(N) = i$ is a limit ordinal and that $T_{\xi(R\text{-simp})}(N') = \oplus\{T_{\xi(M)}(N') \mid M \in R\text{-simp}\}$ for all $N' \in R\text{-mod}$ with $k(N') < i$. If $x \in T_{\xi(R\text{-simp})}(N)$, then for some $j < i$, we have $x \in F_j(N) \cap T_{\xi(R\text{-simp})}(N) = \oplus\ \{T_{\xi(M)}(F_j(N)) \mid M \in R\text{-simp}\} \subseteq \oplus\{T_{\xi(M)}(N) \mid M \in R\text{-simp}\}$. The reverse containment is trivial and so we have equality.

(1) ⇒ (5): Let $I \in \mathcal{L}_{\xi(R\text{-simp})}$. By (1), $R/I = \oplus\{T_{\xi(M)}(R/I) \mid M \in R\text{-simp}\}$. Since R/I is cyclic, only a finite number of summands in the above decomposition are nonzero, say those corresponding to $M_1,\ldots,M_n \in R\text{-simp}$. For $1 \leq i \leq n$, let K_i be the $\xi(M_i)$-purification of I in R. Then $R/I = \oplus K_i/I$. For $1 \leq i \leq n$, set $I_i = \sum_{j\neq i} K_j$. Then these I_i clearly satisfy conditions (i) - (iii).

Now suppose that $I = \bigcap_{i=1}^{m} I'_i$ where $I'_i \in \mathcal{L}_{\xi(M'_i)}$ for $M'_i \in R\text{-simp}$, and that conditions (ii) and (iii) are satisfied for the M'_i. Since $I \subseteq I'_i$ for each i and since

R/I_i' is $\xi(M_i')$-torsion, R/I has a simple homomorphic image isomorphic to M_i'. By (3), R/I, therefore, has a submodule isomorphic to M_i' and so $T_{\xi(M_i')}(R/I) \neq 0$. Therefore, $\{M_1',\ldots,M_m'\} \subseteq \{M_1,\ldots,M_n\}$.

Assume that $m < n$. Then there exists an M_j such that $R/I_i' \notin \mathcal{L}_{\xi(M_j)}$ for $1 \leq i \leq m$. Pick $x + I \in T_{\xi(M_j)}(R/I)$. Then there exists an i, $1 \leq i \leq m$, with $x \notin I_i'$ and so $[Rx + I_i']/I_i'$ is a homomorphic image of $[Rx + I]/I$ that belongs to $\mathcal{T}_{\xi(M_i')}$. But $M_i' = M_k$ for some $k \neq j$ and hence $\mathcal{T}_{\xi(M_j)} \cap \mathcal{T}_{\xi(M_k)} \neq \{0\}$, which is a contradiction. Therefore $m = n$.

(5) $\Rightarrow$ (6): Let $\mathcal{A}$ be the set defined on the right-hand side of the equality in (6). By (5), $\mathcal{L}_{\xi(R\text{-simp})} \subseteq \mathcal{A}$. Conversely, assume that $I \in \mathcal{A} \smallsetminus \mathcal{L}_{\xi(R\text{-simp})}$. Replacing I by its $\xi(R\text{-simp})$-purification if necessary, we can in fact assume that $I \in \mathcal{C}_{\xi(R\text{-simp})}$. By Proposition 12.4, $I \in \mathcal{C}_{\xi(M)}$ for every $M \in R\text{-simp}$. Since $I \in \mathcal{A}$, there exist left ideals $I_1,\ldots,I_n$ of R, $I_i \in \mathcal{L}_{\xi(M_i)}$, with $I \supseteq \cap I_i$. We then have a canonical R-epimorphism $\nu\colon R/[\cap I_i] \to R/I$ and a canonical R-monomorphism $\lambda\colon R/[\cap I_i] \to \oplus R/I_i$. There therefore exists an R-homomorphism β making the diagram

$$\begin{array}{ccccc} 0 & \longrightarrow & R/[\cap I_i] & \xrightarrow{\ \lambda\ } & \oplus R/I_i \\ & & \Big\downarrow \nu & & \Big\downarrow \beta \\ 0 & \longrightarrow & R/I & \longrightarrow & E(R/I) \end{array}$$

commute. Since $\nu \neq 0$, then β is also nonzero and so there exists an index i for which $(R/I_i)\beta \neq 0$. But R/I_i is $\xi(M_i)$-torsion and $E(R/I)$ is $\xi(M_i)$-torsion-free since R/I is $\xi(M_i)$-torsion-free. From this contradiction we deduce (6).

(6) ⇒ (3): Let $I \in \mathcal{L}_{\xi(R\text{-simp})}$. By (6), there exist $M_1,\ldots,M_n \in R$-simp and $I_i \in \mathcal{L}_{\xi(M_i)}$ $(1 \leq i \leq n)$ such that $I \supseteq \cap I_i$. Without loss of generality, we can assume that n is minimal with this property.

Let $M \in R$-simp and assume that we have an R-epimorphism $\alpha\colon R/I \to M$. To prove (3) we have to show that R/I has a submodule isomorphic to M.

Let $x = (1 + I)\alpha$. If $I_1x = 0$ then M is $\xi(M_1)$-torsion and so $M = M_1$. Since $T_{\xi(M_1)}(R/I) \neq 0$, there is then a submodule of R/I isomorphic to M_1 and hence to M. Therefore, assume that $I_1x \neq 0$. In particular, pick $r_1 \in I_1$ such that $r_1x \neq 0$. If $(I_2:r_1)r_1x = 0$ then M is $\xi(M_2)$-torsion and, as before, we are done. Therefore, assume that $(I_2:r_1)r_1x \neq 0$. Continue in this manner. If none of $M_1,\ldots,M_{n-1}$ are isomorphic to M, then we have obtained $r_2,\ldots,r_{n-1}$ with $r_i \in (I_i:r_{i-1} \cdot \ldots \cdot r_1)$ such that $r_ir_{i-1} \cdot \ldots \cdot r_1x \neq 0$. Then $(I_n:r_{n-1} \cdot \ldots \cdot r_1) \in \mathcal{L}_{\xi(M_n)}$ and $(I_n:r_{n-1} \cdot \ldots \cdot r_1)r_{n-1} \cdot \ldots \cdot r_1 \subseteq \cap I_i \subseteq I$. Therefore $(I_n:r_{n-1} \cdot \ldots \cdot r_1)r_{n-1} \cdot \ldots \cdot r_1x = 0$ and so

$M \cong M_n$. Since $T_{\xi(M_n)}(R/I) \neq 0$, R/I has a submodule isomorphic to M and we are done. □

A torsion theory $\tau \in R\text{-tors}$ is said to be <u>completely reducible</u> if and only if for every $\tau' \leq \tau$ there exists a $\tau'' \leq \tau$ for which $\tau = \tau' \oplus \tau''$.

(23.7) PROPOSITION: <u>The following conditions are equivalent for</u> $\tau \in R\text{-tors}$:

(1) τ <u>is completely reducible</u>.

(2) τ <u>is semisimple and has an atomic direct decomposition</u>.

PROOF: (1) ⇒ (2): Let $\tau' = \xi(R\text{-simp} \cap \mathcal{J}_\tau)$. Then $\tau' \leq \tau$. By (1), there exists a $\tau'' \leq \tau$ with $\tau = \tau' \oplus \tau''$. If $\tau'' \neq \xi$ then by Proposition 12.1 there exists an $M' \in R\text{-simp}$ with $\xi(M') \leq \tau''$, whence $M' \in \mathcal{J}_\tau$. But in that case, $M' \in R\text{-simp} \cap \mathcal{J}_\tau \subseteq \mathcal{J}_{\tau'}$, which is a contradiction. Therefore, we must have $\tau'' = \xi$ and so $\tau = \tau'$, proving that τ is semisimple.

Now let $M \in R\text{-simp} \cap \mathcal{J}_\tau$. By hypothesis there exists a $\tau'' \leq \tau$ with $\tau = \xi(M) \oplus \tau''$. Since $\tau = \xi(R\text{-simp} \cap \mathcal{J}_\tau)$, we must have $\tau'' = \xi([R\text{-simp} \cap \mathcal{J}_\tau] \setminus \{M\})$. By Proposition 23.5, we then have $\tau = \oplus\{\xi(M) \mid M \in R\text{-simp} \cap \mathcal{J}_\tau\}$, proving (2).

(2) ⇒ (1): Let $\tau' \leq \tau$ and let $\tau'' = \vee\{\xi(M) \mid M \in R\text{-simp} \cap [\mathcal{J}_\tau \setminus \mathcal{J}_{\tau'}]\}$. Then by (2), $\tau = \tau' \oplus \tau''$. □

(23.8) PROPOSITION: The following conditions are equivalent for a ring R:

(1) Every $\tau \in R$-tors is centrally splitting.

(2) Every $\tau \in R$-tors is completely reducible.

(3) R is left semiartinian and χ has an atomic direct decomposition.

(4) R is left semiartinian and every $\tau \in R$-prop is stable.

(5) Every $\tau \in R$-tors has an atomic direct decomposition.

(6) R is isomorphic to a finite direct product of left semiartinian left local rings.

PROOF: (1) ⇒ (6): By Proposition 23.1, (1) implies that χ is completely reducible and so, by Proposition 23.7, χ is semisimple. This implies that R is left semiartinian.

Pick $M_0 \in R$-simp. By (1) there then exists a $\tau' \in R$-tors with $\chi = \xi(M_0) \oplus \tau'$. In particular, $R = T_{\xi(M_0)}(R) \oplus T_{\tau'}(R)$. If R is $\xi(M_0)$-faithful, then $R = T_{\tau'}(R)$ and $\xi(M_0) = \xi$, which is a contradiction. Therefore, $T_{\xi(M_0)}(R) \neq 0$ and so R has a left ideal isomorphic to M_0.

Since $T_{\xi(M_0)}(R)$ and $T_{\tau'}(R)$ are both two-sided ideals of R, we in fact have the ring direct product $R = T_{\xi(M_0)}(R) \times T_{\tau'}(R)$. Let $R_1 = T_{\xi(M_0)}(R)$ and

$S_1 = T_{\tau'}(R)$. By Proposition 22.16, every $\sigma \in R_1$-tors is centrally splitting and so, by the same argument as above, R_1 is left semiartinian. Moreover, R_1 is left local since any simple left R_1-module is simple as a left R-module and so is isomorphic to M_0.

Now consider S_1. Again by Proposition 22.16, every $\sigma \in S_1$-tors is centrally splitting and so S_1 is left semiartinian. Let M_1 be a simple left S_1-module. Then S_1 contains an isomorphic copy of M_1. Since M_1 is also a simple left R-module, it cannot be isomorphic to M_0 in R-mod. Therefore, we can repeat the preceding process to get $S_1 = R_2 \times S_2$ where $R_2 = T_{\xi(M_1)}(R)$ and $S_2 = T_{\tau'}(R)$ for some $\tau' \in$ R-tors.

Continuing in this manner, we obtain, for each positive integer n, a decomposition $R = R_1 \times R_2 \times \ldots \times R_n \times S_n$ of R where

(i) $R_i = T_{\xi(M_i)}(R)$ for some $M_i \in$ R-simp;

(ii) if $i \neq j$ then $M_i \neq M_j$; and

(iii) each R_i is left semiartinian and left local.

We want to show that for some positive integer n, we have $S_n = 0$.

For each positive integer n, set $I_n = \times_{i=1}^{n} R_i$ and consider the ascending chain $I_1 \subseteq I_2 \subseteq \ldots$ of two-sided ideals of R. Let $I = \bigcup_{i=1}^{\infty} I_i$. Since each I_i is an idempotent two-sided ideal of R, so is I. Therefore, if $\tau = \xi(R/I)$ we have $\tau \in$ R-jans with $L(\tau) = I$. By (1), τ is centrally splitting and so $\chi = \tau \oplus \tau''$ for some $\tau'' \in$ R-tors. By Proposition 22.12, $R = T_\tau(R) + I$. By Proposition 22.10, I is τ^*-torsion and so, by Proposition 22.12, I is τ-torsion-free. Therefore, $R = T_\tau(R) \oplus I$ and so, in particular, we can write $1 = e_1 + e_2$ where $e_1 \in T_\tau(R)$ and $e_2 \in I$. By the definition of I, there exists a positive integer k with $e_2 \in I_k$. Then for all $j > k$, we have $e_2 T_{\xi(R/I_j)}(R) = 0$. Since e_2 is the identity on I, this implies that $I_k = I_{k+1} = \dots$ and so $R = R_1 \times \dots \times R_k$, where each R_i is a left semiartinian left local ring.

(6) ⇒ (5): Suppose that $R = R_1 \times \dots \times R_k$ where each R_i is left semiartinian and left local. Then for every left R-module M, $M = \oplus R_i M$. By Proposition 12.13, R_i-tors $= \{\xi, \chi\}$ for $1 \leq i \leq k$ and so $\xi(R_i M)$ must be atomic in R-mod. This clearly implies (5).

(5) ⇒ (4): Let N be a nonzero left R-module. By (5), $\xi(N) = \oplus\{\xi(M) \mid M \in \mathcal{A}\}$ for some subset $\mathcal{A}$ of R-simp. Therefore, $T_{\xi(M)}(N) \neq 0$ for some $M \in \mathcal{A}$ and so N contains a submodule isomorphic to M. Since this is true for all N, R is left semiartinian.

To prove the second part of (4) it suffices, by

Proposition 12.11, to prove that $\chi(M)$ is stable for every $M \in$ R-simp. In particular, let $N \in \mathcal{J}_{\chi(M)}$ and let $E = E(N)$. Since χ has an atomic direct decomposition, then $E = \oplus\{T_{\xi(M')}(E) \mid M' \in \text{R-simp}\}$. But $T_{\xi(M)}(E) = 0$ lest $T_{\xi(M)}(E) \cap N \neq 0$ and so $E = \oplus\{T_{\xi(M')}(E) \mid M' \neq M\} \in \mathcal{J}_{\chi(M)}$.

(4) $\Rightarrow$ (3): Let N be an injective left R-module and let $N' = \oplus\{T_{\xi(M)}(N) \mid M \in \text{R-simp}\}$. We first claim that N' is also injective. Indeed, assume that it is not injective and let $E(N') \subseteq N$. Then $E(N')/N' \neq 0$ and so by (4) there exists a submodule N_0 of $E(N')$ with $N_0/N' \cong M$ for some $M \in$ R-simp. Since $\xi(M)$ is stable, $T_{\xi(M)}(N)$ is a direct summand of N and so $N = T_{\xi(M)}(N) \oplus N_1$ where N_1 contains $N_2 = \oplus\{T_{\xi(M')}(N) \mid M' \neq M\}$ as a large submodule. Then $N_1 = T_{\chi(M)}(N)$ and so $M \cong N_0/N' \cong N_1/N_2$ is $\chi(M)$-torsion, which is a contradiction. Therefore, N' is injective and so $N = N' \oplus N''$. If $N'' \neq 0$ then there exists a submodule M of N'' isomorphic to an element of R-simp. But then $M \subseteq N'$, which is a contradiction. Therefore, $N'' = 0$ and so $N = \oplus\{T_{\xi(M)}(N) \mid M \in \text{R-simp}\}$.

Now let N be an arbitrary left R-module. Then if $E = E(N)$, by the above argument we have $E = \oplus\{T_{\xi(M)}(E) \mid M \in \text{R-simp}\}$. Let $x \in N$. Then $x = \Sigma x_M$ where each $x_M \in T_{\xi(M)}(E) \cap N = T_{\xi(M)}(N)$. Therefore, $x \in \oplus\{T_{\xi(M)}(N) \mid M \in \text{R-simp}\}$. This implies that

$N = \oplus\{T_{\xi(M)}(N) \mid M \in R\text{-simp}\}$ for every left R-module N and so $\chi = \oplus\{\xi(M) \mid M \in R\text{-simp}\}$, proving (3).

(3) ⇒ (2): Let $\tau \in R\text{-tors}$. Since R is left semiartinian, τ is semisimple by Proposition 12.6. Indeed, $\tau = \xi(R\text{-simp} \cap \mathcal{J}_\tau)$ and so τ has at atomic direct decomposition since χ has. Then (2) follows by Proposition 23.7.

(2) ⇒ (1): Since χ is completely reducible, for each $\tau \in R\text{-tors}$ there exists a $\tau' \in R\text{-tors}$ with $\tau \oplus \tau' = \chi$. By Proposition 23.1, τ is therefore centrally splitting. □

(23.9) PROPOSITION: <u>The following conditions are equivalent for a ring</u> R:

(1) $T_\tau(\_)$ <u>is exact for every</u> $\tau \in R\text{-tors}$.

(2) R <u>is isomorphic to a finite direct product of right perfect left local rings</u>.

PROOF: (1) ⇒ (2): By Proposition 12.9, R is left semiartinian and so $\chi = \xi(R\text{-simp})$. Now let N be a left R-module and let $N' = \oplus\{T_{\xi(M)}(N) \mid M \in R\text{-simp}\}$. Then we have the exact sequence of left R-modules $0 \to N' \to N \to N/N' \to 0$. By (1), for any $M \in R\text{-simp}$ the functor $T_{\xi(M)}(\_)$ is exact and so it induces the exact sequence $0 \to T_{\xi(M)}(N') \to T_{\xi(M)}(N) \to T_{\xi(M)}(N/N') \to 0$. But $T_{\xi(M)}(N') = T_{\xi(M)}(N)$ and so $T_{\xi(M)}(N/N') = 0$ for all $M \in R\text{-simp}$. Since R is left semiartinian, this implies that $N/N' = 0$ and so $N = N'$, which proves that χ has an atomic direct decomposition. By

Proposition 23.8, R is then isomorphic to a finite direct product of left semiartinian left local rings R_i. Moreover, each R_i is of the form $T_{\xi(M)}(R)$ for some $M \in$ R-simp. Also, by Proposition 12.9, for each i we have $J(R_i)$ being right T-nilpotent and $R_i/J(R_i)$ left semiartinian.

To prove (2) it therefore suffices to show that each R_i is right perfect. To do this it suffices, by [9, Theorem P], to show that each $R_i/J(R_i)$ is completely reducible.

Let $\nu_i: R_i \to R_i/J(R_i)$ be the canonical ring surjection and let $H_i = [\mathrm{soc}(R_i/J(R_i))]\nu_i^{-1}$. Then H_i is a two-sided ideal of R_i and we are done if we can show that $H_i = R_i$. Assume that they are not equal. If K is a maximal left ideal of R_i containing H_i and if $R_i/K \cong R_i/K'$ for some left ideal K' of R, then it follows that $K' \supseteq H_i$. Since R_i is left semiartinian, $H_i \supset J(R_i)$ and so, since $J(R_i)$ is the intersection of all maximal left ideals of R_i, it follows that there exists a maximal left ideal K of R_i that does not contain H_i. Then there exists a maximal left ideal K' of R_i containing H_i so that $R_i/K \not\cong R_i/K'$, contradicting the fact that R_i is left local. Therefore, we must have $R_i = H_i$.

(2) ⇒ (1): By Proposition 23.8, R is left semiartinian. Also, for every $\tau \in$ R-tors, each R_i is either τ-torsion or τ-torsion-free and so $\mathfrak{F}_\tau$ is clearly closed

under taking homomorphic images. Therefore (1) follows from Proposition 5.5. □

References for Section 23

Bronowitz [21, 22]; Teply [154].

CHAPTER VI

REPRESENTATION THEORY

24. TOPOLOGIES ON SUBSETS OF R-prop

For a ring R, we want to define topologies on subsets of R-prop that are compatible with the lattice structure on R-prop. There are several methods of constructing such topologies, one of which is the following: If U is a subset of R-tors, then by Proposition 8.3, we have $\text{gen}(\vee U) = \cap\{\text{gen}(\tau) \mid \tau \in U\}$. Therefore, if U is closed under taking finite joins and if X is a nonempty subset of R-prop, U induces a topology on X a base for which is $\{\text{gen}(\tau) \cap X \mid \tau \in U\}$. We call this the U-order topology on X. Note that, in particular, if X = R-sp, then the basic open sets of the U-order topology are precisely $\{\text{pgen}(\tau) \mid \tau \in U\}$.

There are two "canonical" choices for U:

(1) U = R-tors. In this case the U-order topology on X is called the full-order topology.

(2) $U = \{\tau \in R\text{-tors} \mid \tau$ is basic$\}$ (refer to

Proposition 8.7). In this case the U-order topology on X is called the basic-order topology.

We can also approach these topologies in another manner. For any R-homomorphism $\alpha: M \to N$, let $D(\alpha) = \{\tau \in R\text{-prop} \mid \mathrm{Hom}(\alpha,E): \mathrm{Hom}_R(N,E) \to \mathrm{Hom}_R(M,E)$ is monic for all $E \in \tau\}$.

(24.1) PROPOSITION: If $\alpha: M \to N$ is an R-homomorphism then $D(\alpha) = \mathrm{gen}(\xi(\mathrm{coker}(\alpha)))$.

PROOF: If E is an injective left R-module, then from the exact sequence $M \overset{\alpha}{\to} N \overset{\eta}{\to} \mathrm{coker}(\alpha) \to 0$ we obtain the exact sequence $0 \to \mathrm{Hom}_R(\mathrm{coker}(\alpha),E) \xrightarrow{\mathrm{Hom}(\eta,E)} \mathrm{Hom}_R(N,E) \xrightarrow{\mathrm{Hom}(\alpha,E)} \mathrm{Hom}_R(M,E)$ and so $\mathrm{Hom}(\alpha,E)$ is monic $\Leftrightarrow \mathrm{Hom}(\eta,E) = 0 \Leftrightarrow E \in \mathcal{F}_{\xi(\mathrm{coker}(\alpha))} \Leftrightarrow \xi(\mathrm{coker}(\alpha)) \leq \chi(E)$, which suffices to prove the result. □

If $\alpha: M \to N$ and $\alpha': M' \to N'$ are R-homomorphisms, then $D(\alpha) \cap D(\alpha') = D(\alpha \oplus \alpha')$. Therefore, if $\mathcal{A}$ is any family of R-homomorphisms closed under taking finite direct sums, then $\{D(\alpha) \cap X \mid \alpha \in \mathcal{A}\}$ is a basis for an order topology on X.

(24.2) PROPOSITION: Let $\tau \in X \subseteq R\text{-prop}$. Then the closure of $\{\tau\}$ in the full-order topology on X is $\mathrm{spcl}(\tau) \cap X$.

PROOF: By definition, τ' belongs to the closure of $\{\tau\}$ if and only if every open neighborhood of τ'

intersects $\{\tau\}$. This clearly happens when $\tau' \leq \tau$. Conversely, if $\tau' \not\leq \tau$ then there exists a left R-module $M \in \mathcal{J}_{\tau'} \setminus \mathcal{J}_{\tau}$. But then $\tau' \in \text{gen}(\xi(M))$ and $\tau \notin \text{gen}(\xi(M))$ and so $\text{gen}(\xi(M)) \cap X$ is an open neighborhood of τ' not containing τ. □

Therefore, the full-order topology on a subset X of R-prop need not, in general, be a T_1-space. However, it is clearly a T_0-space.

(24.3) PROPOSITION: <u>Let</u> $\tau \in X \subseteq$ R-prop. <u>Then</u> $\text{gen}(\tau) \cap X$ <u>is quasicompact in the full order topology on</u> X.

PROOF: If $\{\text{gen}(\tau_i) \cap X\}$ is an open cover of $\text{gen}(\tau) \cap X$, then there exists an index k with $\tau \in \text{gen}(\tau_k) \cap X$ and so $\text{gen}(\tau) \cap X = \text{gen}(\tau_k) \cap X$. □

We now turn to the special case in which we are most interested, that of $X =$ R-sp.

(24.4) PROPOSITION: <u>For a semiprime torsion theory</u> $\tau \in$ R-prop, <u>the following conditions are equivalent</u>:

(1) $\tau = \wedge U$ <u>for some finite</u> $U \subseteq$ R-sp.

(2) $\text{pgen}(\tau)$ <u>is quasicompact in any order topology on</u> R-sp.

(3) $\text{pgen}(\tau)$ <u>is quasicompact in the full-order topology on</u> R-sp.

PROOF: (1) ⇒ (2): Let $\{\text{pgen}(\tau') \mid \tau' \in V\}$ be an open cover of pgen(τ). Then there exists a finite subset V' of V such that $\cup\{\text{pgen}(\tau'') \mid \tau'' \in U\} \subseteq \cup\{\text{pgen}(\tau') \mid \tau' \in V'\}$. But $\tau = \wedge U$ implies that $\text{pgen}(\tau) = \cup\{\text{pgen}(\tau'') \mid \tau'' \in U\}$ by Proposition 19.19 and so (2) follows.

(2) ⇒ (3): The proof is trivial.

(3) ⇒ (1): Since $\{\text{pgen}(\tau') \mid \tau' \in \text{pgen}(\tau)\}$ is an open cover of pgen(τ) in the full-order topology, there exists a finite subset U of pgen(τ) with $\text{pgen}(\tau) = \cup\{\text{pgen}(\tau'') \mid \tau'' \in U\}$. Then $\tau = \sqrt{\tau} = \wedge\{\sqrt{\tau''} \mid \tau'' \in U\} = \wedge U$. □

(24.5) PROPOSITION: _Let_ R _be a left noetherian ring over which every_ $\tau \in$ R-sp _is stable. Then every subset of_ R-sp _that is open in the full-order topology on_ R-sp _is of the form_ pgen(τ) _for some_ $\tau \in$ R-prop.

PROOF: Let U be a subset of R-sp that is open in the full-order topology. By Propositions 19.18 and 1.6, there exists a $\tau \in$ R-prop such that $\mathcal{J}_\tau = \{M \in \text{R-mod} \mid \text{supp}(M) \cap U = \emptyset\}$. Let $\tau' \in U$. Then $\tau' = \chi(E)$ for some indecomposable injective left R-module E. Moreover, $\text{ass}(E) = \{\tau'\}$ and so $\tau' \in U \cap \text{supp}(E)$. Thus E is not τ-torsion and so $\tau \in \text{supp}(E)$. By Proposition 21.14, this means that $\tau \leq \tau'$ and so $\tau' \in \text{pgen}(\tau)$. Thus $U \subseteq \text{pgen}(\tau)$.

Conversely, assume that $\tau' \in \text{pgen}(\tau)$, where $\tau' = \chi(E)$ for some indecomposable injective left R-module E. Then $\text{ass}(E) \subseteq \text{pgen}(\tau)$. By Proposition 21.3, E is τ-torsion-free. Therefore, there exists a $\tau'' \in \text{supp}(E) \cap U$. By Proposition 21.14, $\tau'' \leq \tau'$ and so by the openness of U, we have $\tau' \in U$. Therefore, $\text{pgen}(\tau) \subseteq U$ and so we have equality. □

It is also worth remarking at this point that, by Proposition 19.16, if R is left semiartinian, then every order topology on R-sp is discrete.

(24.6) PROPOSITION: <u>If</u> R <u>is commutative, then</u> R-sp <u>with the basic order topology is homeomorphic to</u> spec(R) <u>with the Zariski topology</u>.

PROOF: Define the function $\theta: \text{spec}(R) \to R\text{-sp}$ by $\theta: P \mapsto \chi(R/P)$. By Corollary 18.8, the function θ is epic. Furthermore, for each prime ideal P of R, we have $\mathcal{T}_{\chi(R/P)} = \{M \in R\text{-mod} \mid \text{for each } m \in M \text{ there exists an } a \in R \setminus P \text{ with } am = 0\}$ (see Chapter VII). This shows that θ must also be monic.

Let $U(I) = \{P \in \text{spec}(R) \mid I \nsubseteq P\}$ be an open subset of spec(R). Then $P \in U(I)$ if and only if $I \in \mathcal{L}_{\chi(R/P)}$, i.e., if and only if $\chi(R/P) \in \text{pgen}(\xi(R/I))$. Therefore, $\theta(U(I)) = \text{pgen}(\xi(R/I))$ and $\theta^{-1}(\text{pgen}(\xi(R/I)) = U(I)$, proving that θ is a homeomorphism. □

If every $\tau \in$ R-prop is semiprime, then we can define topologies on R-sp other than the order topologies.

(24.7) PROPOSITION: *If every* $\tau \in$ R-prop *is semiprime then the function* $U \mapsto \text{pgen}(\wedge U)$ *is a closure operator on* R-sp.

PROOF: Clearly, $\text{pgen}(\wedge\emptyset) = \emptyset$. Also, from the definition, $U \subseteq \text{pgen}(\wedge U)$ for any subset U of R-sp. Since every $\tau \in$ R-prop is semiprime, $\wedge U = \sqrt{\wedge U}$ and so $\text{pgen}(\wedge U) = \text{pgen}(\sqrt{\wedge U}) = \text{pgen}(\wedge\text{pgen}(\wedge U))$. Finally, by Propositions 8.4 and 19.19, $\text{pgen}(\wedge(U_1 \cup U_2)) = \text{pgen}((\wedge U_1) \wedge (\wedge U_2)) = \text{pgen}(\wedge U_1) \cup \text{pgen}(\wedge U_2)$. □

The closure operator defined in Proposition 24.7 yields a topology on R-sp, for suitable R, the open sets of which are precisely those of the form $\text{R-sp} \setminus \text{pgen}(\tau)$ for $\tau \in$ R-tors. We call this the *reverse-order topology* on R-sp.

(24.8) COROLLARY: *Let* R *be a ring over which every* $\tau \in$ R-prop *is semiprime. Then for any left* R-*module* M, supp(M) *is open in the reverse-order topology on* R-sp.

PROOF: It is straightforward to check that $\text{supp}(M) = \text{R-sp} \setminus \text{pgen}(\tau)$ where $\tau = \wedge[\text{R-sp} \setminus \text{supp}(M)]$. □

(24.9) PROPOSITION: *If* $\tau \in$ R-perf *then* R_τ-sp *is homeomorphic to* pgen(τ) *in the order or in the reverse-order topologies (when the latter exists).*

PROOF: By Corollary 19.25, $\hat{\tau}_{\#}$: pgen(τ) $\to$ R_τ-sp is bijective. That it is a homeomorphism for either topology follows by Proposition 9.1(3). □

A closed subset of a topological space is *irreducible* if and only if it is not the union of two proper closed subsets.

(24.10) PROPOSITION: *Let* R *be a left seminoetherian ring. Then the following conditions are equivalent for a subset* U *of* R-sp:

(1) U *is irreducible.*

(2) U = pgen(τ) *for some* $\tau \in$ R-sp.

PROOF: (1) $\Rightarrow$ (2): If U is irreducible then U = pgen($\wedge$U). If there exist $\tau_1, \tau_2 \in$ gen($\wedge$U) with $\tau_1 \wedge \tau_2 = \wedge U$, then U = pgen($\wedge$U) = pgen($\tau_1$) $\cup$ pgen(τ_2) by Proposition 19.19. By irreducibility, we must have pgen($\wedge$U) = pgen(τ_i) for $i = 1$ or $i = 2$ and so $\wedge U = \tau_i$. By Proposition 20.8, this implies that $\wedge U \in$ R-sp.

(2) $\Rightarrow$ (1): Suppose that U = pgen(τ) for $\tau \in$ R-sp and that $U = U_1 \cup U_2$ where the U_i = pgen($\wedge U_i$). Then U = pgen($\wedge U_1$) $\cup$ pgen($\wedge U_2$) = pgen($[\wedge U_1] \wedge [\wedge U_2]$). Since $\tau \in U$, this means that $[\wedge U_1] \wedge [\wedge U_2] \leq \tau$ and so, by Proposition 20.8, $\wedge U_i \leq \tau$ for $i = 1$ or $i = 2$, whence $U = U_i$. □

References for Section 24

Golan [55, 56]; Hacque [69]; Marot [99, 100].

25. REPRESENTATION OF RINGS OVER R-sp

The purpose of this section is to introduce a method for the representation of rings by sheaves analogous to the one used by algebraic geometers in commutative ring theory.

If $\tau \leq \tau' \in$ R-prop, then for every left R-module M there exists, by Proposition 7.3, a unique R-homomorphism $\alpha_{\tau,\tau'}(M)$ making the following diagram commute.

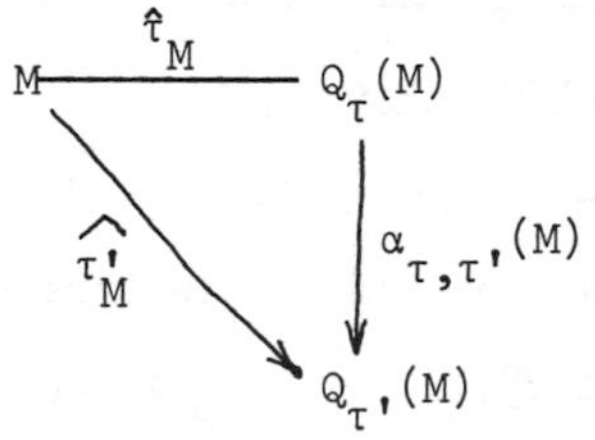

Now let X be a subset of R-prop and fix a topology on X. Let open(X) be the category the objects of which are the subsets of X open in the given topology and the morphisms of which are the inclusion maps $\lambda_{UV}: U \to V$. Define the functor $\overline{Q}$: open$(X)^{op} \times$ R-mod $\to$ R-mod as follows:

(1) If U is an object of open(X) and if M is a left R-module, then $\overline{Q}(U,M) = Q_{\wedge U}(M)$.

(2) If $\lambda_{UV}: U \to V$ is a morphism in open(X) and if $\varphi: M \to N$ is a morphism in R-mod, then

$\bar{Q}(\lambda_{UV},\varphi)\colon \bar{Q}(V,M) \to \bar{Q}(U,N)$ is defined by the commutative diagram

$$\begin{array}{ccc} \bar{Q}(V,M) & \xrightarrow{\alpha_{\wedge V,\wedge U}(M)} & \bar{Q}(U,M) \\ {\scriptstyle Q_{\wedge V}(\varphi)}\downarrow & & \downarrow{\scriptstyle Q_{\wedge U}(\varphi)} \\ \bar{Q}(V,N) & \xrightarrow{\alpha_{\wedge V,\wedge U}(N)} & \bar{Q}(U,N) \end{array}$$

(25.1) PROPOSITION: *Let* X *be a subset of* R-prop *and let* M *be a left* R-*module. Then the functor* $\bar{Q}(\_,M)\colon \text{open}(X)^{op} \to$ R-mod *is a separated presheaf.*

PROOF: Let U be an object of open(X) and let $\{U_i\}$ be an open cover of U. For each i let $\lambda_i\colon U_i \to U$ be the inclusion map. Then we have to show that the R-homomorphism $\psi\colon \bar{Q}(U,M) \to \Pi\bar{Q}(U_i,M)$ defined by $q \mapsto \langle q\bar{Q}(\lambda_i,M)\rangle$ is monic. But $\ker(\psi) = \{q \in \bar{Q}(U,M) \mid q\bar{Q}(\lambda_i,M) = 0 \text{ for all } i\} = \bigcap_i\{T_{\wedge U_i}(Q_{\wedge U}(M))\} = T_{\bigwedge_i(\wedge U_i)}(Q_{\wedge U}(M))$. Since $\bigwedge_i(\wedge U_i) = \wedge U$ by Proposition 8.4, we then have $\ker(\psi) = 0$. □

Since $\bar{Q}(\_,M)$ is separated, we have a canonical embedding $\bar{Q}(\_,M) \to \tilde{Q}(\_,M)$ of $\bar{Q}(\_,M)$ into its associated sheaf. In particular, we note that $\tilde{Q}(\_,R)$ is a sheaf of rings and that for every left R-module M, $\tilde{Q}(\_,M)$ is a sheaf of $\tilde{Q}(\_,R)$-modules (i.e., for every open subset U of X, we have that $\tilde{Q}(U,M)$ is a left $\tilde{Q}(U,R)$-module).

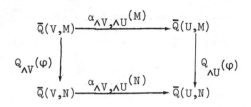

The natural choice for X is R-sp or the set of all minimal elements of R-sp. The natural choice of a topology on X is some topology for which the stalk of $\tilde{Q}(\_,M)$ at any $\tau \in X$ is just $Q_\tau(M)$. The full order topology clearly satisfies this property. We then obtain a ringed space $(X,\tilde{Q}(\_,R))$ which we want to use to represent R. That is to say, we want to show that "local" results about the stalks $Q_\tau(M)$ can be blown up into "global" results about the modules M. If X contains all the (minimal prime) torsion theories of the form $\chi(N)$ for $N \in$ R-simp, then we have the additional property that $\tilde{Q}(X,M) = Q_{\wedge X}(M) = Q_\xi(M) = M$ for any left R-module M, by Proposition 12.3.

We shall limit ourselves to the considerations of the cases $X =$ R-sp and $X = \{\chi(N) \mid N \in \text{R-simp}\}$.

(25.2) PROPOSITION: <u>The following conditions are equivalent for a left</u> R-<u>module</u> M:

(1) $M = 0$.

(2) $Q_\tau(M) = 0$ <u>for all</u> $\tau \in$ R-sp.

(3) $Q_{\chi(N)}(M) = 0$ <u>for all</u> $N \in$ R-simp.

PROOF: (1) ⇒ (2) ⇒ (3): The proof is trivial.

(3) ⇒ (1): If $Q_{\chi(N)}(M) = 0$ for all $N \in$ R-simp, then $M \in \mathcal{T}_{\wedge\{\chi(N) \mid N \in \text{R-simp}\}}$ and so $M = 0$ by Proposition 12.3. □

(25.3) PROPOSITION: *Let* $\alpha: M \to M'$ *be an R-homomorphism. Then the following conditions are equivalent:*

(1) α *is an R-monomorphism.*

(2) $Q_\tau(\alpha)$ *is an* R_τ*-monomorphism for all* $\tau \in$ R-sp.

(3) $Q_{\chi(N)}(\alpha)$ *is an* $R_{\chi(N)}$*-monomorphism for all* $N \in$ R-simp.

PROOF: If $K = \ker(\alpha)$ then $\ker(Q_\tau(\alpha)) = Q_\tau(K)$ for any torsion theory τ by the left exactness of $Q_\tau(\_)$. Then the result follows from Proposition 25.2. □

(25.4) PROPOSITION: *Let* $\alpha: M \to M'$ *be an R-homomorphism for which* $Q_{\chi(N)}(\alpha): Q_{\chi(N)}(M) \to Q_{\chi(N)}(M')$ *is an* $R_{\chi(N)}$*-epimorphism for all* $N \in$ R-simp. *Then* α *is an R-epimorphism.*

PROOF: Let $\nu: M' \to M'/M\alpha$ be the canonical surjection. For every $N \in$ R-simp, consider the commutative diagram

$$\begin{array}{ccc}
M & \xrightarrow{\widehat{\chi(N)}_M} & Q_{\chi(N)}(M) \\
\downarrow{\scriptstyle \alpha} & & \downarrow{\scriptstyle Q_{\chi(N)}(\alpha)} \\
M' & \xrightarrow{\widehat{\chi(N)}_{M'}} & Q_{\chi(N)}(M') \\
\downarrow{\scriptstyle \nu} & & \downarrow{\scriptstyle Q_{\chi(N)}(\nu)} \\
M'/M\alpha & \xrightarrow{\widehat{\chi(N)}_{M'/M\alpha}} & Q_{\chi(N)}(M'/M\alpha)
\end{array}$$

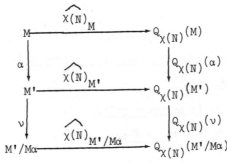

Since $\alpha\nu = 0$, then by Proposition 25.2, $0 = Q_{\chi(N)}(\alpha)Q_{\chi(N)}(\nu)$ and so, since $Q_{\chi(N)}(\alpha)$ is an $R_{\chi(N)}$-epimorphism, we have $Q_{\chi(N)}(\nu) = 0$. Therefore, $\text{im}(\widehat{\chi(N)}_{M'/M\alpha}) \subseteq \text{im}(Q_{\chi(N)}(\nu)) = 0$ and so $Q_{\chi(N)}(M'/M\alpha) = 0$ for every $N \in R\text{-simp}$. By Proposition 25.2, this implies that $M'/M\alpha = 0$, proving that α is an epimorphism. □

(25.5) COROLLARY: Let R be a ring over which every $\tau \in R\text{-sp}$ is exact. Then the following conditions are equivalent for an R-homomorphism $\alpha: M \to M'$:

(1) α is an R-epimorphism.

(2) $Q_\tau(\alpha)$ is an R_τ-epimorphism for all $\tau \in R\text{-sp}$.

(3) $Q_{\chi(N)}(\alpha)$ is an $R_{\chi(N)}$-epimorphism for all $N \in R\text{-simp}$.

PROOF: The proof of (1) ⇒ (2) follows from the hypothesis and the proof of (2) ⇒ (3) is trivial. (3) ⇒ (1) follows from Proposition 25.4. □

(25.6) PROPOSITION: Let $\alpha: M \to M'$ be an R-homomorphism. Then the following conditions are equivalent:

(1) α is an isomorphism.

(2) $Q_\tau(\alpha)$ is an isomorphism for all $\tau \in R\text{-sp}$.

(3) $Q_{\chi(N)}(\alpha)$ is an isomorphism for all $N \in R\text{-simp}$.

PROOF: The proof of (1) ⇒ (2) ⇒ (3) is trivial and (3) ⇒ (1) follows from Propositions 25.3 and 25.4. □

(25.7) PROPOSITION: Let R be a ring for which R-simp is finite. Let M be a left R-module and suppose that for each $N \in$ R-simp $Q_{\chi(N)}(M)$ has one of the following properties:

(1) It is noetherian.

(2) It is artinian.

(3) It is finitely-generated.

Then M has the same property.

PROOF: (1) Let $M_1 \subseteq M_2 \subseteq \cdots$ be an ascending chain of submodules of M. If $N \in$ R-simp, then $Q_{\chi(N)}(M_1) \subseteq Q_{\chi(N)}(M_2) \subseteq \cdots$ is an ascending chain of submodules of $Q_{\chi(N)}(M)$ and so there exists a natural number $k(N)$ such that $Q_{\chi(N)}(M_{k(N)}) = Q_{\chi(N)}(M_{k(N)+j})$ for every natural number j. Since R-simp is finite, there exists a natural number k such that $k \geq k(N)$ for every $N \in$ R-simp. Let j be a natural number and let $\lambda\colon M_k \to M_{k+j}$ be the inclusion map. Then $Q_{\chi(N)}(\lambda)$ is an isomorphism for every $N \in$ R-simp and so, by Proposition 25.6, λ is an isomorphism. Therefore M is noetherian.

(2) This is proven similarly.

(3) Since R-simp is finite and $Q_{\chi(N)}(M)$ is finitely generated for every $N \in$ R-simp, there exists a finite subset

A of M such that $\{m\widehat{\chi(N)}_M \mid m \in A\}$ generates $Q_{\chi(N)}(M)$ for every $N \in$ R-simp. Let M' be the submodule of M generated by the elements of A and let $\lambda: M' \to M$ be the inclusion map. Then $Q_{\chi(N)}(\lambda)$ is an isomorphism for all $N \in$ R-simp and so, by Proposition 25.6, λ is an isomorphism. Therefore, M is finitely generated. □

<u>References for Section 25</u>

Golan [55]; Marot [99, 100]; Sim [143, 144].

CHAPTER VII

EXAMPLES

EXAMPLE 1. The "ancestor" of the torsion theories discussed in this volume is the following: Let R be a commutative integral domain. A left R-module M is said to be torsion-free if and only if $0 \neq r \in R$ and $0 \neq m \in M$ imply that $0 \neq rm$. It is easily seen that the left R-modules satisfying this condition form a class $\mathcal{F}_\tau$ for some $\tau \in$ R-tors. Moreover, τ is faithful, stable, and perfect. If every finitely generated ideal of R is principal, then the torsion-free left R-modules are precisely the flat left R-modules [89, p. 135].

EXAMPLE 2. Let D be the set of all two-sided non zero divisors of a ring R and let $\mathcal{L} = \{{}_R I \subseteq R \mid I \cap D \neq \emptyset\}$. Then $\mathcal{L}$ is an idempotent filter if and only if R is a left Öre ring, i.e., if and only if for every $a \in R$ and $d \in D$ there exist $a' \in R$ and $d' \in D$, respectively, such that $d'a = a'd$. In this case, $\mathcal{L}$ defines a torsion theory $\mu(D) \in$ R-tors called the classical torsion theory on

R-mod. The ring $R_{\mu(D)}$ is called the <u>classical ring of left quotients</u> of R; it satisfies the following conditions:

(1) If $d \in D$ then d^{-1} exists in $R_{\mu(D)}$;

(2) For each $q \in R_{\mu(D)}$ there exist $a \in R$ and $d \in D$ with $q = d^{-1}a$. In particular, every $I \in \mathcal{L}$ contains an invertible element of $R_{\mu(D)}$ and so $\mu(D)$ is perfect.

See also [162].

<u>EXAMPLE 3 [60]</u>. Let R be an arbitrary ring and let A be a multiplicatively closed subset of R containing 1. Then $\{M \in R\text{-mod} \mid \text{for each } m \in M \text{ there exists an } a \in A \text{ with } am = 0\} = \mathcal{J}_\tau$ for some $\tau \in R$-tors. We denote this τ by $\mu(A)$. (Note that Example 2 is then a special case of this example.)

PROPOSITION: <u>If</u> R <u>is commutative and</u> P <u>is a prime ideal of</u> R, <u>then</u> $\mu(R \setminus P) = \chi(R/P)$.

PROOF: If $a \in R \setminus P$, then $ab \in P$ implies that $b \in P$. Therefore, $T_{\mu(R \setminus P)}(R/P) = 0$ and so $\mu(R \setminus P) \leq \chi(R/P)$. Now let $I \notin \mathcal{L}_{\mu(R \setminus P)}$. Then I does not contain any element of $R \setminus P$ whence $I \subseteq P$. But then I annihilates R/P and so $I \notin \mathcal{L}_{\chi(R/P)}$. Therefore, $\mathcal{L}_{\chi(R/P)} \subseteq \mathcal{L}_{\mu(R \setminus P)}$ and so $\chi(R/P) \leq \mu(R \setminus P)$, proving equality. □

If we take the special case $\mathcal{A} = \{I\}$, then $\mathcal{J}_\tau$ consists of all left R-modules each element of which is annihilated by a power of I. Such a τ is called the <u>I-adic torsion theory</u>. A left R-module M is absolutely pure with respect to this theory if and only if the natural map $\theta: M \to \mathrm{Hom}_R(I,M)$ defined by $\theta(m): a \mapsto am$ is bijective.

We note that some form of finiteness condition on the members of $\mathcal{A}$ is necessary. To see this consider, the following example, due to M. Teply. Let K be a field and let $R = K[x_1, x_2,\ldots]$ be the polynomial ring over K in countably many commuting indeterminates. Let I be the maximal ideal of R generated by $x_1, x_2,\ldots$ and let $\mathcal{A} = \{I\}$. Set $\mathcal{J} = \{M \in R\text{-mod} \mid$ every element of M is annihilated by a power of $I\}$. Let $H = \sum_{j=1}^{\infty} I^j x_j \subseteq I$. If $y \in I$ and $n = \max\{j \mid x_j$ appears in $y\}$ then $I^n y \subseteq H$. Therefore $I/H \in \mathcal{J}$. Clearly, $R/I \in \mathcal{J}$ and so if $\mathcal{J}$ were equal to $\mathcal{J}_\tau$ for some $\tau \in R\text{-tors}$ we would have $R/H \in \mathcal{J}$ by Proposition 1.6. But this implies that $I^n \subseteq H$ for some natural number n, which is a contradiction since for every natural number n, $x_n^n \in I^n \setminus H$.

Now suppose that the ring R is left noetherian and that I is a two-sided ideal of R. The ring R is said to have the <u>Artin-Rees Property</u> with respect to I if and only if

<u>EXAMPLE 4</u>. Let R be a left noetherian ring and let P be a two-sided prime ideal of R. Then $C(P) = \{a \in R \mid ab \in P$ implies that $b \in P\}$ is a multiplicatively closed set containing 1. Moreover,

$$(*) \qquad \mu(C(P)) = \chi(R/P)$$

and this torsion theory is prime. Considerable research has been done on these theories, see [60, 61, 70, 83, 94, 107].

The relation (*) remains true if P is assumed to be only semiprime. Such theories have been studied in [81, 82, 83, 95].

<u>EXAMPLE 5</u>. Let $\mathcal{A}$ be a family of two-sided ideals of R that are finitely generated as left R-modules and let $\mathcal{J} = \{M \in R\text{-mod} \mid$ for each $m \in M$ there exists a finite product of members of $\mathcal{A}$ contained in $(0:m)\}$. Then $\mathcal{J} = \mathcal{J}_\tau$ for some $\tau \in R$-tors.

In particular, if $R = \mathbb{Z}$ and if $\mathcal{A}$ is a set of prime ideals of $\mathbb{Z}$ then there exists a $\tau \in \mathbb{Z}$-tors with $\mathcal{J}_\tau = \{M \in \mathbb{Z}\text{-mod} \mid$ each $m \in M$ is annihilated by a finite product of members of $\mathcal{A}\}$. Indeed every nontrivial $\tau \in \mathbb{Z}$-prop is of this form for, given such a τ, let $\mathcal{A}$ be the set of all prime ideals generated by the prime integers in the prime factorization of the generators of the ideals of $\mathcal{L}_\tau$. [104].

A multiplicatively closed subset A of R is said to be a <u>left Öre set</u> if and only if for every $a \in A$ and $r \in R$ there exist $a' \in A$ and $r' \in R$ satisfying $r'a = a'r$.

PROPOSITION: The <u>following conditions are equivalent for a multiplicatively closed subset</u> A <u>of a ring</u> R:

(1) A <u>is a left Öre set</u>.

(2) <u>For every left</u> R-<u>module</u> M, $T_{\mu(A)}(M) = \{m \in M \mid (0:m) \cap A \neq 0\}$.

(3) <u>For every</u> $a \in A$, $Ra \in \mathcal{L}_{\mu(A)}$.

PROOF: (1) ⇒ (2): One can straightforwardly show from (1) that $\{m \in M \mid (0:m) \cap A = \emptyset\}$ is a submodule of M and so is contained in $T_{\mu(A)}(M)$. The reverse implication is immediate.

(2) ⇒ (3): Let $a \in A$. Then $a(1 + Ra) = 0$ and so, by (2), $1 + Ra \in T_{\mu(A)}(R/Ra)$. Hence Ra is $\mu(A)$-dense in R.

(3) ⇒ (1): Let $a \in A$ and $r \in R$. Since R/Ra is $\mu(A)$-torsion, there exists an $a' \in A$ such that $a'(1 + Ra) = 0$. Thus $a'r = r'a$ for some $r' \in R$. □

If $a \in A$, then one can show [52, p. 415] that $\widehat{\mu(A)}(a)^{-1}$ exists in $R_{\mu(A)}$. Also, for any such A, $\mu(A) \in R$-perf.

for every submodule N of a finitely generated left R-module M and for every non-negative integer n there exists an integer h(n) for which $I^{h(n)}M \cap N \subseteq I^nN$.

As a consequence of Proposition 11.4 we then have the following result.

PROPOSITION: *The following conditions are equivalent for a two-sided ideal* I *of a left noetherian ring* R:

(1) R *has the Artin-Rees property with respect to* I.

(2) *The* I-*adic torsion theory on* R-mod *is stable*.

EXAMPLE 6. If W is a flat right R-module, then $\{M \in R\text{-mod} \mid W \otimes_R M = 0\}$ is $\mathcal{J}_\tau$ for some $\tau \in R$-tors. Indeed, $\tau = \chi(\mathrm{Hom}(W, \mathbb{Q}/\mathbb{Z}))$. Then $\mathcal{L}_\tau = \{{}_RI \subseteq R \mid WI = W\}$. Moreover, if W is projective as a right R-module, then we define the *trace* of W by $H = \Sigma\{\alpha W \mid \alpha \in \mathrm{Hom}_R(W, R_R)\}$ and note that H is a two-sided ideal of R. By the Dual Basis Theorem, WH = W and $H^2 = H$. Therefore, τ is jansian. Moreover, in this case, R_τ is isomorphic to the double centralizer of W [34, Theorem 2.1].

If W is finitely generated and if S is the endomorphism ring of W, then the following conditions are equivalent [33]:

(1) τ is perfect.

(2) ${}_S W$ is finitely generated and projective.

(3) $\mathrm{Hom}_S(W,\_)$: S-mod $\to$ R-mod is exact and commutes with direct sums.

EXAMPLE 7. A left R-module M is said to be nonsingular if and only if no element of M is annihilated by a large left ideal of R. The class of all nonsingular left R-modules is $\mathcal{F}_\tau$ for some torsion theory $\tau \in$ R-tors. This torsion theory is called the Goldie torsion theory and is denoted by τ_G. Indeed, $\tau_G = \xi(\{M/N \mid N$ is a large submodule of $M\})$. Clearly, τ_G is stable. If τ_G is faithful, then $\tau_G = \chi(R)$ but this is not true in general. Moreover, if τ_G is faithful then, using Proposition 1.7, one sees that $\mathcal{L}_{\tau_G}$ consists precisely of the large left ideals of R. Therefore $L(\tau_G) = \mathrm{soc}(R)$. If soc(R) is large in R, as in the case of left semiartinian rings, this implies that τ_G is jansian. If R is a commutative integral domain, then τ_G is the torsion theory defined in Example 1. Moreover, the τ_G-injective left R-modules are just the injective left R-modules.

The Goldie torsion theory has been investigated extensively by several authors. See [5, 29, 32, 54, 59, 65, 153, 154, 156, 157].

EXAMPLE 8 [39]. Let $\mathcal{A} = \{{}_R I \subseteq R \mid$ for all $a \in R \setminus I$ there exists a $b \in R$ such that $(I:ba)$ is a proper large left ideal of $R\}$. Then $\mathcal{A} = \mathcal{L}_\tau$ for some $\tau \in R$-tors. Similarly, let $\mathcal{A}' = \{{}_R I \subseteq R \mid$ for all $a \in R \setminus I$ there is no $0 \neq b \in R$ such that $(I:a)b = 0\}$. Then $\mathcal{A}' = \mathcal{L}_{\tau'}$ for some $\tau' \in R$-tors. Moreover, the following conditions are equivalent:

(1) $\tau = \tau'$.

(2) $I \in \mathcal{L}_\tau$ if and only if I is a large left ideal of R.

(3) $I \in \mathcal{L}_{\tau'}$ if and only if I is a large left ideal of R.

(4) ${}_R R$ is nonsingular.

EXAMPLE 9. The torsion theory $\chi(R)$ is often called the Lambek torsion theory. It was first studied in [161]. The ring $R_{\chi(R)}$ is called the maximal ring of left quotients of R and has been described extensively in [89]. A result of Johnson [89, p. 106] asserts that $R_{\chi(R)}$ is regular if and only if the Goldie torsion theory τ_G is faithful and, as we have observed previously, in this case $\chi(R) = \tau_G$. If τ_G is faithful, then the Lambek torsion theory is perfect if and only if $R_{\chi(R)}$ is completely reducible [136, Theorem 1.6]; it is exact if and only if every $I \in \mathcal{L}_{X(R)}$ is projective [165, Proposition 3.1].

<u>EXAMPLE 10 [64]</u>. The class of all projective completely reducible left R-modules certainly satisfies the conditions of Proposition 1.6(2) and so is $\mathcal{J}_\tau$ for some $\tau \in$ R-tors. Thus we have a topology $X_\tau({}_RR)$ on R a basis of neighborhoods of 0 in which is the family of all left ideals I such that R/I is projective and completely reducible. Then R is discrete under this topology if and only if R is a completely reducible ring. Moreover, we have the following generalization of the Wedderburn-Artin structure theorem.

PROPOSITION: <u>The following conditions on a ring</u> R <u>are equivalent</u>:

(1) <u>The topology</u> $X_\tau(R)$ <u>is complete</u>.

(2) R <u>is isomorphic to a cartesian product of endomorphism rings of vector spaces over division rings</u>.

<u>EXAMPLE 11 [5, p. 40]</u>. Let K be a field and let R = K[x,y] be the polynomial ring over K in two commuting indeterminates x and y. Let I = (x,y). Then I is a maximal ideal of R, and the I-adic torsion theory τ on R-mod (see Example 5) has the property that $R_\tau = R$.

EXAMPLE 12 [18]. Let R be either left coherent or finitely generated over its center. Then $\{{}_R I \subseteq R \mid R/I$ has the descending chain condition on finitely generated submodules$\}$ is an idempotent filter that defines a torsion theory on R-mod called the *Björk torsion theory*.

EXAMPLE 13 [1]. Let R be an arbitrary ring. A left R-module M is *flat cohereditary* if and only if M/N is flat for every submodule N of M. The class of all flat cohereditary left R-modules is $\mathcal{T}_\tau$ for some $\tau \in$ R-tors.

EXAMPLE 14 [86]. Let $M \in$ R-mod. A left R-module N is said to be M-*distinguished* if and only if for every nonzero R-homomorphism $\alpha\colon N' \to N$ there exists an R-homomorphism $\beta\colon M \to N'$ such that $\beta\alpha \neq 0$. Then $\{N \in \text{R-mod} \mid N$ is M-distinguished$\}$ is $\mathcal{F}_\tau$ for some $\tau \in$ R-tors. Indeed, $\tau = \chi(M')$ where M' is a direct sum of a complete set of representatives of the isomorphism classes of M-distinguished epimorphic images of M.

If $S = \mathrm{Hom}_R(M,M)$, then $N \in \mathcal{F}_\tau$ if and only if the R-homomorphism $E(N) \to \mathrm{Hom}_S(\mathrm{Hom}_R(M,R), \mathrm{Hom}_R(M,E(N)))$ defined by $x \mapsto \varphi_x$ where $(m)[(\alpha)\varphi_x] = (m\alpha)x$ for each $\alpha \in \mathrm{Hom}_R(M,R)$ is an isomorphism.

EXAMPLE 15 [167]. A module $M \in R\text{-mod}$ is <u>semiprime</u> if and only if, for each $0 \neq m \in M$, there exists an $\alpha \in \text{Hom}_R(M,R)$ for which $(m\alpha)m \neq 0$. The class of semiprime left R-modules is seen immediately to be closed under taking submodules and direct products. Assume that R is left self-injective. If $M \in R\text{-mod}$ is semiprime and $0 \neq x \in E(M)$ then there exists an $r \in R$ with $0 \neq rx \in M$. By the semiprimeness of M there exists an $\alpha \in \text{Hom}_R(M,R)$ for which $(rx)\alpha(rx) \neq 0$. By the injectivity of R there exists an R-homomorphism $\beta: E(M) \to R$ the restriction of which to M is given by $m \mapsto (m\alpha)r$. Then $0 \neq (rx)\beta x = r[(x\beta)x]$, and so $(x\beta)x \neq 0$. Therefore $E(M)$ is semiprime. By Proposition 1.4, there then exists a $\tau \in R\text{-tors}$ such that $\mathcal{F}_\tau$ is the class of all semiprime left R-modules.

EXAMPLE 16 [88]. Let R be a hereditary noetherian prime ring. Define the subfunctor T of the identity functor $R\text{-mod} \to R\text{-mod}$ by $T(M) = \{m \in M \mid Im = 0$ for some invertible two-sided ideal I of $R\}$ and $T(\alpha) =$ restriction of α to $T(\text{domain } \alpha)$. Then T is a torsion functor and so determines a torsion theory $\tau \in R\text{-tors}$. Moreover, $\mathcal{L}_\tau = \{I \subseteq R \mid I$ is a large left ideal of R containing an invertible two-sided ideal of $R\}$.

Let $\mathcal{L}' = \{{}_R I \subseteq R \mid$ every submodule of every factor module of R/I belongs to $\mathcal{F}_\tau\}$. Then $\mathcal{L}'$ is an idempotent filter and so determines a torsion theory $\tau' \in$ R-tors.

EXAMPLE 17 [118]. A left R-module M is called an S-module if and only if every homomorphic image of E(M) is injective. If R is left noetherian, then the collection of all S-modules is $\mathcal{J}_\tau$ for some $\tau \in$ R-tors. If R is left artinian, then τ is jansian.

EXAMPLE 18 [138]. Let R be the ring of all real-valued continuous functions on the unit interval [0,1] and let $\mathcal{L}$ be the set of all ideals H of R such that $\{x \in [0,1] \mid f(x) \neq 0$ for all $f \in H\}$ is an open dense subset of [0,1]. Then $\mathcal{L} = \mathcal{L}_{\chi(R)}$. Moreover, there is no flat right R-module W such that $I \in \mathcal{L}$ if and only if WI = W.

EXAMPLE 19 [124]. Let R be a left hereditary ring and define the subfunctor T of the identity functor R-mod → R-mod by $T(M) = M \cap J(E(M))$ and $T(\alpha)$ = restriction of α to T(domain α). Then T is a torsion functor and so defines a torsion theory $\tau \in$ R-tors. Moreover, $\tau = \chi$ if and only if J(M) = M for every injective left R-module M and $\tau = \xi$ if and only if R is a left V-ring (see [44]).

EXAMPLE 20 [145]. Let P be a projective generator of R-mod and let $\mathcal{A}$ be a set of quotient objects of P which contains P. An exact sequence $0 \to M' \to M \to M'' \to 0$ is said to be $\mathcal{A}$-pure if and only if $\mathrm{Hom}_R(N,M) \to \mathrm{Hom}_R(N,M'') \to 0$ is exact for every $N \in \mathcal{A}$. Let $\tau = \xi(\mathcal{A} \setminus \{P\})$. Then $M'' \in \mathcal{F}_\tau$ if and only if every sequence $0 \to M' \to M \to M'' \to 0$ is $\mathcal{A}$-pure.

EXAMPLE 21 [25]. Let R be a commutative ring and let $M \in R$-mod. Let $M[X]$ be the module of formal polynomials with coefficients in M. Such a polynomial can be evaluated at any $r \in R$. Moreover, there exists a $\tau \in R$-tors for which $\mathcal{F}_\tau = \{M \in R\text{-mod} \mid$ for every $0 \neq f \in M[X]$ there exists an $r \in R$ with $f(r) \neq 0\}$.

EXAMPLE 22 [17]. Let A be a subset of a ring R and let $\mathcal{L} = \{{}_RI \subseteq R \mid$ for any sequence $\langle a_i \rangle$ of elements of A and for any $r \in R$ there exists an $n \geq 1$ such that $a_n a_{n-1} \cdot \ldots \cdot a_1 r \in I\}$. Then $\mathcal{L}$ is an idempotent filter and so $\mathcal{L} = \mathcal{L}_\tau$ for some $\tau \in R$-tors. We denote this torsion theory by $\upsilon(A)$.

PROPOSITION: For a left ideal I of R,

(1) $\upsilon(I) \leq \xi(R/I)$.

(2) If I is two sided, then $\upsilon(I) = \xi(R/I)$.

(3) <u>If</u> I <u>is two-sided then</u> $\upsilon(I) = \chi$ <u>if and only if</u> I <u>is right</u> T-<u>nilpotent</u>.

PROOF: (1) Let ${}_R H \notin \mathcal{L}_{\xi(R/I)}$. Then there exists an $a_1 \in I$ with $(H:a_1) \notin \mathcal{L}_{\xi(R/I)}$. Similarly there exists an $a_2 \in I$ with $(H:a_2a_1) = ((H:a_1):a_2) \notin \mathcal{L}_{\xi(R/I)}$. Continue in this manner to construct a sequence $\langle a_i \rangle$ of elements of I with $(H:a_n a_{n-1} \cdot \ldots \cdot a_1) \notin \mathcal{L}_{\xi(R/I)}$ for all $n \geq 1$. But then $H \notin \mathcal{L}_{\upsilon(I)}$. Thus $\upsilon(I) \leq \xi(R/I)$.

(2) If I is two sided, then $I \in \mathcal{L}_{\upsilon(I)}$ and so $\xi(R/I) \leq \upsilon(I)$, whence we have equality.

(3) $\upsilon(I) = \chi \Leftrightarrow 0 \in \mathcal{L}_{\upsilon(I)} \Leftrightarrow I$ is right T-nilpotent. □

Also note that for $A' \subseteq A$ we have $\upsilon(A) \leq \upsilon(A')$.

<u>EXAMPLE 23 [15, 102]</u>. Let R be the ring of all 2×2 upper triangular matrices over a field K and let $e = \begin{bmatrix} 0 & 0 \\ 0 & 1 \end{bmatrix}$. Then $I = ReR$ is an idempotent two-sided ideal of R and so determines a jansian torsion theory $\tau \in R$-tors. Since ${}_R I$ is a direct summand of R, R/I is projective as a left R-module and so, by Proposition 22.12, $\mathcal{F}_\tau$ is closed under taking homomorphic images. However, R/I is not flat as a right R-module and so, by Proposition 22.10, τ is not centrally splitting.

EXAMPLE 24. A left R-module is said to be a QF-3' module if and only if $\{N \in R\text{-mod} \mid \mathrm{Hom}_R(N,M) = 0\} = \mathbb{J}_{\chi(M)}$. In particular, this condition is equivalent to $\mathrm{Hom}_R(M,R) = 0 \Rightarrow \mathrm{Hom}_R(M',R) = 0$ for all submodules M' of M. If ${}_R R$ is a left QF-3' module, we say that R is a left QF-3' ring. This is true if and only if $T_{\chi(R)}(M) = \cap\{\ker(\alpha) \mid \alpha \in \mathrm{Hom}_R(M,R)\}$ for every left R-module M. See [16, 80, 85, 163].

EXAMPLE 25 [G. Bergman]. Let G be a nondiscrete ordered group in which the identity e is not the infimum of any countable set of elements greater than e (e.g., G is an uncountable product of copies of $\mathbb{Z}$ ordered lexicographically). Let S be the semigroup $\{g \in G \mid g > e\}$. Let K be a field and let R be the semigroup algebra of K over S. Let I be the ideal of R generated by all $g \in S$. Then I is an idempotent maximal ideal of R and so there exists a $\tau \in R$-jans with $\mathcal{L}_\tau = \{I,R\}$. Then τ is not of finite type since I is not finitely generated. On the other hand, let $H_1 \subseteq H_2 \subseteq \dots$ be a countably infinite ascending chain of proper left ideals of R different from I. Then for each i there exists an $e \neq g_i \in S \setminus H_i$. Since $e \neq \inf\{g_i\}$, there is a $g \in S$ satisfying $e \neq g < g_i$ for all i. Then $g \notin \cup H_i$ and so $\cup H_i \notin \mathcal{L}_\tau$. Therefore τ is noetherian.

EXAMPLE 26 [144]. Let F be a field and let R be the subring of F_3 consisting of all elements of the form

$$A = \begin{bmatrix} a_{11} & 0 & 0 \\ a_{21} & a_{22} & 0 \\ a_{31} & a_{32} & a_{11} \end{bmatrix}$$

Let $I = \{A \in R \mid a_{11} = 0\}$ and $H = \{A \in R \mid a_{22} = 0\}$. Then I and H are two-sided ideals of R that are maximal as left ideals. Let $\tau = \chi(R/I)$. The τ is prime since R/I is simple. Moreover, R is left artinian and so $\mathcal{L}_\tau$ has a minimal element, proving that τ is jansian. In fact, $L(\tau) = H$. Since $HT_\tau(R) = 0$, it then follows that $T_\tau(R) = 0$ and so τ is faithful.

Let

$$x = \begin{bmatrix} 0 & 0 & 0 \\ 0 & 0 & 0 \\ 0 & 1 & 0 \end{bmatrix} \in H.$$

One can show that if $q \in R_\tau$ then

$$xq = \begin{bmatrix} 0 & 0 & 0 \\ 0 & 0 & 0 \\ a_{31} & a_{32} & 0 \end{bmatrix}$$

for $a_{31}, a_{32} \in F$. Therefore, if $h \in H$,

$$xqh = \begin{bmatrix} 0 & 0 & 0 \\ 0 & 0 & 0 \\ a_{31} & 0 & 0 \end{bmatrix}$$

for some $a_{31} \in F$. Hence τ cannot be perfect, for if τ were perfect we would have $R_\tau H = R_\tau$ and hence there would exist $\{q_i\} \subseteq R_\tau$ and $\{h_i\} \subseteq H$ with $\Sigma q_i h_i = 1$ whence $\Sigma x q_i h_i = x$, which we have seen cannot happen.

But R is left noetherian, so we do have $\tau \in R$-fin. Therefore τ is not exact.

<u>EXAMPLE 27 [130, 131]</u>. Call a left ideal I of R <u>weakly large</u> in R if and only if for any $a_1, \ldots, a_n \in R$ there exists a $0 \neq r \in R$ with $ra_i \in I$ for all i. A ring is called a <u>left</u> ATF-<u>ring</u> if and only if there exists a $\tau \in R$-tors with $\mathcal{L}_\tau = \{{}_R I \subseteq R \mid I \text{ is weakly large in } R\}$. Indeed in this case τ must be equal to $\chi(R)$. If R is a left ATF-ring, then R is prime and the center of R is an integral domain. If R is finitely generated as a module over its center, then R is a left ATF-ring if and only if it is prime. Furthermore, if R is a left ATF-ring, then any subring of $R_{\chi(R)}$ containing R is also a left ATF-ring.

<u>EXAMPLE 28 [143]</u>. If R is a nonzero (noncommutative) integral domain that is not a left Öre

domain, then there are no $\chi(R)$-cocritical modules. Indeed, if there were a $\chi(R)$-cocritical module M, then by Proposition 18.2 there would exist a $\chi(R)$-cocritical left ideal I of R. Let $a, b \in R \setminus \{0\}$. Then for all $0 \neq c \in I$, we have $Iac \cap Ibc \neq 0$ by Proposition 18.2. Thus there exist $a', b' \in I$ such that $b'ac = a'bc$ whence $b'a = a'b$, contradicting the fact that R is not a left Ore domain.

Note also that in this example, 0 is a prime ideal of R but $\chi(R) = \chi(R/0)$ is not a prime torsion theory.

<u>EXAMPLE 29</u>. Let F be a field, x_1 and x_2 be indeterminates over F, and let $R = F[x_1,x_2]$. Define $I = Rx_1 + Rx_2^2$. Then direct computation shows that R/I is a uniform left R-module but is not cocritical.

<u>EXAMPLE 30 [M. Teply]</u>. Let R be the ring of all matrices of the form

$$\begin{bmatrix} a_{11} & a_{12} & a_{13} \\ 0 & a_{22} & a_{23} \\ 0 & 0 & a_{33} \end{bmatrix}$$

where $a_{11} \in \mathbb{Z}$ and the other $a_{ij} \in \mathbb{Q}$. Define $\tau_1, \tau_2 \in$ R-tors by

$$\mathcal{L}_{\tau_1} = \left\{ {}_R I \subseteq R \;\middle|\; \begin{bmatrix} a_{11} & a_{12} & a_{13} \\ 0 & 0 & 0 \\ 0 & 0 & 0 \end{bmatrix} \in I \right.$$

for all $a_{11} \in \mathbb{Z}$ and $a_{12}, a_{13} \in \mathbb{Q}\}$

$$\text{and} \quad \mathcal{L}_{\tau_2} = \left\{ {}_R I \subseteq R \;\middle|\; \begin{bmatrix} a_{11} & a_{12} & a_{13} \\ 0 & a_{22} & a_{23} \\ 0 & 0 & 0 \end{bmatrix} \in I \right.$$

for all $a_{11} \in \mathbb{Z}$ and the other $a_{ij} \in \mathbb{Q}\}$.

Then we have the following:

(1) τ_i is faithful $(i = 1, 2)$.

(2) τ_i is stable $(i = 1, 2)$.

(3) Every τ_2-torsion module is injective.

(4) $$R_{\tau_1} = R_{\tau_2} = \left\{ \begin{bmatrix} a_{11} & a_{12} & a_{13} \\ 0 & a_{22} & a_{23} \\ 0 & a_{32} & a_{33} \end{bmatrix} \;\middle|\; a_{11} \in \mathbb{Z}, \right.$$

all other $a_{ij} \in \mathbb{Q}\}$.

(5) τ_1 is not noetherian and τ_2 is noetherian.

(6) Every $I \in \mathcal{L}_{\tau_2}$ is projective.

<u>EXAMPLE 31 [164]</u>. Let $R = \mathbb{Z}^{\mathbb{N}}$ and let $H = \mathbb{Z}^{(\mathbb{N})}$.
Then H is an idempotent two-sided ideal of R and so H

defines $\tau \in$ R-jans. Since H is projective as a left ideal, then by Proposition 16.3, τ is exact. But τ is not noetherian and so is not perfect.

<u>EXAMPLE 32 [21]</u>. Let K be a field and let $R = K^{\mathbb{N}}$. Set $I_1 = \{\langle a_i \rangle \in R \mid a_{2i+1} = 0 \text{ for all } i \in \mathbb{N}\}$ and $I_2 = \{\langle a_i \rangle \in R \mid a_{2i} = 0 \text{ for all } i \in \mathbb{N} \text{ not a power of } 2\}$. Let $I = I_1 \cap I_2$. Then I, I_1, and I_2 are idempotent two-sided ideals of R and so define jansian torsion theories τ, τ_1, and τ_2, respectively. Moreover, $\tau = \tau_1 \oplus \tau_2$ and so τ is directly decomposable.

On the other hand, τ is not semisimple. Indeed, $I \in \mathcal{L}_\tau$ implies that $I + K^{(\mathbb{N})} \in \mathcal{L}_\tau$. But $soc(R/[I + K^{(\mathbb{N})}]) = 0$ so that if τ were semisimple we would have to have $R = I + K^{(\mathbb{N})}$, which is false.

<u>EXAMPLE 33 [11]</u>. Let V be a vector space over a field F of countably infinite dimension and let R be the ring of endomorphisms of V. Let I be the ideal of R consisting of all endomorphisms of V of finite rank. Then I is in fact the only nonzero proper ideal of R. Indeed, I is a prime ideal of R. Moreover, $\chi(R)$ and $\chi(R/I)$ are both coatoms of R-tors.

<u>EXAMPLE 34</u>. If $\tau \in$ R-tors, then $\mathcal{E}_\tau$ is an abelian category. When is this category equivalent to R'-mod for some ring R'? Several results have been obtained in this area, among them the following.

PROPOSITION [151]: <u>Let</u> U $\in$ R-mod <u>and let</u> $S = \mathrm{Hom}_R(U,U)$. <u>Then the following conditions are equivalent</u>:

(1) U <u>is a generator of</u> R-mod.

(2) <u>If</u> V <u>is an injective cogenerator of</u> R-mod <u>and</u> $W = \mathrm{Hom}_R(U,V) \in$ S-mod, <u>then</u> R-mod <u>is categorically equivalent to</u> $\mathcal{E}_{\chi(W)}$.

<u>EXAMPLE 35 [51]</u>. If $\tau \in$ R-tors, then $\mathcal{J}_\tau$ is an additive subcategory of R-mod. Suppose we have (additive) category equivalences $F: \mathcal{J}_\tau \to$ R'-mod and G: R'-mod $\to \mathcal{J}_\tau$ for some ring R'. Then

(1) F and G yield isomorphisms

$$\mathrm{Hom}_R(M,M') \to \mathrm{Hom}_{R'}(F(M),F(M')) \text{ and}$$

$$\mathrm{Hom}_{R'}(N,N') \to \mathrm{Hom}_R(G(M),G(M'))$$

for all $M, M' \in \mathcal{J}_\tau$ and all N, N' $\in$ R'-mod.

(2) If $U = G(R') \in$ R-mod-R', then $F \cong \mathrm{Hom}_R(U,\_)$ and $G \cong U \otimes_{R'} -$.

(3) R' is canonically isomorphic to $\mathrm{Hom}_R(U,U)$.

(4) U generates $\mathcal{J}_\tau$.

(5) $_RU$ is finitely generated, is quasiprojective, and generates each of its submodules.

(6) $U_{R'}$ is faithfully flat.

EXAMPLE 36 [89]. Let $\mathcal{A}$ be a full abelian subcategory of R-mod which generates R-mod. Let $\mathcal{A}'$ [respectively, $\mathcal{A}''$] be the class of all $M \in$ R-mod for which $\mathrm{Hom}_R(\_,M)\colon \mathcal{A}^{op} \to \mathbb{Z}$-mod preserves monomorphisms [respectively, kernels]. Then there exists a $\tau \in$ R-tors for which $\mathcal{F}_\tau = \mathcal{A}'$. Moreover, if $\mathcal{A}$ is closed under taking kernels and all modules in $\mathcal{A}$ are projective, then $\mathcal{E}_\tau = \mathcal{A}''$. In [89] this torsion theory is used to prove that every small abelian category admits a faithful and full exact functor into the full subcategory $\mathcal{E}_\tau$ of R-mod for some ring R and some $\tau \in$ R-tors.

INDEX OF NOTATION

INDEX OF TERMINOLOGY

REFERENCES

1. H. S. Ahluwalia, A study of flat modules, Ph.D. thesis, Indiana University, 1971.

2. Toma Albu, Modules de torsion à support fini, C. R. Acad. Sc. Paris 273 (1971), A335-A338.

3. J. S. Alin, Primary decomposition of modules, Math. Z. 107 (1968), 319-325.

4. J. S. Alin and E. P. Armendariz, TTF-classes over perfect rings, J. Austral. Math. Soc. 11 (1970), 499-503.

5. J. S. Alin and S. E. Dickson, Goldie's torsion theory and its derived functor, Pacific J. Math. 24 (1968), 195-203.

6. E. P. Armendariz, Quasi-injective modules and stable torsion theories, Pacific J. Math. 31 (1969), 277-280.

7. G. Azumaya, Completely faithful modules and self-injective rings, Nagoya Math. J. 27 (1966), 697-708.

8. G. Azumaya, Some properties of TTF-classes, in "Proceedings of the conference on orders, group rings, and related topics", Lecture Notes in Mathematics 353, Springer-Verlag, Berlin, 1973.

9. H. Bass, Finitistic dimension and homological generalization of semiprimary rings, Trans. Amer. Math. Soc. 95 (1960), 466-488.

10. J. Beachy, A characterization of torsion free modules over rings of quotients, Proc. Amer. Math. Soc. 34 (1972), 15-19.

11. J. Beachy, On maximal torsion radicals, Canad. J. Math. 25 (1973), 712-726.

12. J. Beachy, On quotient functors which preserve small projectives, preprint 1972.

13. R. L. Bernhard, Splitting hereditary torsion theories over semiperfect rings, Proc. Amer. Math. Soc. 22 (1969), 681-687.

14. R. L. Bernhard, On splitting in hereditary torsion theories, Pacific J. Math. 39 (1971), 31-38.

15. R. L. Bernhard, On centrally splitting, preprint 1973.

16. L. Bican, QF-3' modules and rings, Comm. Math. Univ. Carolinae 14 (1973), 295-303.

17. L. Bican, P. Jambor, T. Kepka, and P. Němec, On rings with trivial torsion parts, Bull. Austral. Math. Soc. 9 (1973), 275-290.

18. J. E. Björk, Radical properties of perfect modules, J. Reine und Ang. Math. 253 (1972), 78-86.

19. P. Bland, Perfect torsion theories, Proc. Amer. Math. Soc. 41 (1973), 349-355.

20. J. J. Bowe, Neat homomorphisms, Pacific J. Math. 40 (1972), 13-21.

21. R. Bronowitz, Decomposition of torsion theories, preprint 1972.

22. R. Bronowitz, Completely reducible torsion classes, preprint 1972.

23. R. Bronowitz and M. Teply, Torsion theories of simple type, preprint 1973.

24. P.-J. Cahen, Premiers et copremiers sur un anneau noethérien, C. R. Acad. Sc. Paris 277 (1973), A277-A280.

25. P.-J. Cahen, Torsion theory and associated primes, Proc. Amer. Math. Soc. 38 (1973), 471-476.

26. M. C. Chamard, Caractérisation de certaines théories de torsion, C. R. Acad. Sc. Paris 271 (1970), A1045-A1048.

27. S. U. Chase, Direct products of modules, Trans. Amer. Math. Soc. 97 (1960), 457-473.

28. A. W. Chatters and A. G. Heinicke, Localisation at a torsion theory in hereditary noetherian rings, Proc. London Math. Soc. 27 (1973), 193-204.

29. T. Cheatham, Finite dimensional torsion free rings, Pacific J. Math. 39 (1971), 113-118.

30. K. L. Chew, Closure operations in the study of rings of quotients, Bull. Math. Soc. Nanyang Univ. (1965), 1-20.

31. R. R. Colby and E. A. Rutter, Jr., Π-flat and Π-projective modules, Arch. Math. 22 (1971), 246-251.

32. R. S. Cunningham, Injective modules over rings of quotients, preprint 1972.

33. R. S. Cunningham, On finite left localizations, preprint 1973.

34. R. S. Cunningham, E. A. Rutter, Jr., and D. T. Turnidge, Rings of quotients of endomorphism rings of projective modules, Pacific J. Math. 41 (1972), 647-668.

35. S. Dickson, Decomposition of modules I. Classical rings, Math. Z. 90 (1965), 9-13.

36. S. Dickson, A torsion theory for abelian categories, Trans. Amer. Math. Soc. 121 (1966), 223-235.

37. S. Dickson, Noetherian splitting rings are artinian, J. London Math. Soc. 42 (1967), 732-736.

38. S. Dickson, Decomposition of modules II. Rings without chain conditions, Math. Z. 104 (1968), 349-357.

39. V. Dlab, The concept of a torsion module, Amer. Math. Monthly 75 (1968), 973-976.

40. V. Dlab, A characterization of perfect rings, Pacific J. Math. 33 (1970), 79-88.

41. V. Dlab, On a class of perfect rings, Canad. J. Math. 22 (1970), 822-826.

42. V. P. Elizarov, General theory of quotient rings, Sibir. Math. Zh. 11 (1970), 526-546. (Russian).

43. C. Faith, Rings with ascending conditions on annihilators, Nagoya Math. J. 27 (1966), 179-191.

44. C. Faith, "Lectures on injective modules and quotient rings," Lecture Notes in Mathematics 49, Springer-Verlag, Berlin, 1967.

45. C. Faith, "Algebra: Rings, modules and categories I," Springer-Verlag, Berlin, 1973.

46. G. D. Findlay, A note on noncommutative localization, J. London Math. Soc. (2) 6 (1972), 39-42.

47. G. D. Findlay and J. Lambek, A generalized ring of quotients. I, Canad. Math. Bull. 1 (1958), 77-85.

48. G. D. Findlay and J. Lambek, A generalized ring of quotients. II, Canad. Math. Bull. 1 (1958), 155-167.

49. L. Fuchs, On quasi-injective modules, Ann. Scuola Norm. Sup. Pisa 23 (1969), 541-546.

50. L. Fuchs, Torsion preradicals and ascending Loewy series of modules, J. Reine und Ang. Math. 239/240 (1969), 169-179.

51. K. R. Fuller, Density and equivalence, preprint 1973.

52. P. Gabriel, Des catégories abéliennes, Bull. Soc. Math. France 90 (1962), 323-448.

53. B. J. Gardner, Rings whose modules form few torsion theories, Bull. Austral. Math. Soc. 4 (1971), 355-359.

54. E. Gentile, Singular submodule and injective hull, Indag. Math. 24 (1962), 426-433.

55. J. Golan, Topologies on the torsion-theoretic spectrum of a noncommutative ring, to appear, Pacific J. Math.

56. J. Golan, On the torsion-theoretic spectrum of a noncommutative ring, submitted for publication, 1973.

57. J. Golan and M. Teply, Torsion free covers, Israel J. Math. 15 (1973), 237-256.

58. J. Golan and M. Teply, Finiteness conditions on filters of left ideals, J. Pure Appl. Algebra 3 (1973), 251-260.

59. A. W. Goldie, Torsion-free modules and rings, J. Algebra 1 (1964), 268-287.

60. A. W. Goldie, Localization in noncommutative noetherian rings, J. Algebra 5 (1967), 89-105.

61. A. W. Goldie, A note on noncommutative localization, J. Algebra 8 (1968), 41-44.

62. A. W. Goldie, The structure of noetherian rings, in "Lectures on Rings and Modules," Lecture Notes in Mathematics 246, Springer-Verlag, Berlin, 1972.

63. O. Goldman, Rings and modules of quotients, J. Algebra 13 (1969), 10-47.

64. O. Goldman, A Weddeburn-Artin-Jacobson structure theorem, preprint 1974.

65. K. R. Goodearl, Singular torsion and splitting properties, Amer. Math. Soc. Memoir 124 (1972).

66. M. Hacque, Localisations exactes et localisations plates, Publ. Dép. Math. Lyon 6-2, 97-117.

67. M. Hacque, Caractérisations générales des localisations, Publ. Dép. Math. Lyon 7-4, 45-103

68. M. Hacque, Une propriete des localisations plates, Publ. Dép. Math. Lyon 7-4, 103-123.

69. M. Hacque, Localisations et schemas affines, Publ. Dép. Math. Lyon 7-2, 1-114.

70. A. G. Heinicke, On the ring of quotients at a prime ideal of a right noetherian ring, Canad. J. Math. 24 (1972), 703-712.

71. G. Helzer, On divisibility and injectivity, Canad. J. Math. 18 (1966), 901-919.

72. K. Hirata, Some types of separable extensions of rings, Nagoya Math. J. 33 (1968), 107-115.

73. A. Hudry, Quelques applications de la localisation au sens de Gabriel, C. R. Acad. Sc. Paris 270 (1970), A8-A10.

74. A. Hudry, Sur la localisation dans une catégorie de modules, C. R. Acad. Sc. Paris 270 (1970), A925-A928.

75. A. Hudry, Sur les anneaux localement homogènes, C. R. Acad. Sc. Paris 271 (1970), A1214-A1217.

76. A. Hudry, Sur les modules primaires de O. Goldman C. R. Acad. Sc. Paris 274 (1972), A1772-A1774.

77. A. Hudry, Une remarque sur la decomposition primaire de O. Goldman, Publ. Dép. Math. Lyon 8-1, 63-68.

78. H. Immediato, Localisation a la manière de Goldie et applications, Publ. Dép. Math. Lyon 7-3, 1-54.

79. J. P. Jans, Some aspects of torsion, Pacific J. Math. 15 (1965), 1249-1259.

80. J. P. Jans, Torsion associated with duality, Tohoku Math. J. 24 (1972), 449-452.

81. A. V. Jategaonkar, Injective modules and classical localization in noetherian rings, Bull. Amer. Math. Soc. 79 (1973), 152-157.

82. A. V. Jategaonkar, Injective modules and localization in noncommutative noetherian rings, preprint 1973.

83. A. V. Jategaonkar, The torsion theory at a semiprime ideal, preprint 1973.

84. R. E. Johnson, Quotient rings of rings with zero singular ideal, Pacific J. Math. 11 (1961), 1385-1392.

85. T. Kato, Torsionless modules, Tohoku Math. J. 20 (1968), 233-242.

86. T. Kato, U-distinguished modules, J. Algebra 25 (1973), 15-24.

87. Y. Kurata, On an n-fold torsion theory in the category ${}_R\mathfrak{M}$, J. Algebra 22 (1972), 559-572.

88. J. Kuzmanovich, Localizations of HNP rings, Trans. Amer. Math. Soc. 173 (1972), 137-157.

89. J. Lambek, "Lectures on rings and modules," Blaisdell, Waltham Mass., 1966.

90. J. Lambek, "Torsion theories, additive semantics, and rings of quotients," Lecture Notes in Mathematics 177, Springer-Verlag, Berlin, 1971.

91. J. Lambek, Bicommutators of nice injectives, J. Algebra 21 (1972), 60-63.

92. J. Lambek, Localization and completion, J. Pure Appl. Algebra 2 (1972), 343-370.

93. J. Lambek, Noncommutative localization, Bull. Amer. Math. Soc. 79 (1973), 857-872.

94. J. Lambek and G. Michler, The torsion theory at a prime ideal of a right noetherian ring, J. Algebra 25 (1973), 364-389.

95. J. Lambek and G. Michler, Localization of right noetherian rings at semiprime ideals, preprint 1973.

96. H. M. Leu and J. J. Hutchinson, Morita contexts and quotient rings, preprint 1973.

97. L. Levy, Torsion-free and divisible modules over non-integral-domains, Canad. J. Math. 15 (1963), 132-151.

98. J. M. Maranda, Injective structures, Trans. Amer. Math. Soc. 110 (1964), 98-135.

99. J. Marot, Faisceau des localisations sur un anneau non nécessairement commutatif, C. R. Acad. Sc. Paris 271 (1970), A1148-A1151.

100. J. Marot, Espace des localisations sur un anneau non nécessairement commutatif, preprint 1970.

101. G. Michler, Goldman's primary decomposition and the tertiary decomposition, J. Algebra 16 (1970), 129-137.

102. R. W. Miller, TTF classes and quasigenerators, preprint 1973.

103. R. W. Miller and D. Turnidge, Co-artinian rings and Morita duality, Israel J. Math. 15 (1973), 12-26.

104. A. Mishina and L. Skornyakov, "Abelian groups and modules," Nauka, Moscow, 1969. (Russian).

105. K. Morita, Localizations in categories of modules. I, Math. Z. 114 (1970), 121-144.

106. K. Morita, Localization in categories of modules. II, J. Reine Angew. Math. 242 (1970), 163-169.

107. K. Morita, Localizations in categories of modules. III, Math. Z. 119 (1971), 313-320.

108. K. Morita, Flat modules, injective modules, and quotient rings, Math. Z. 120 (1971), 25-40.

109. K. Morita, Quotient rings in "Ring Theory" (R. Gordon ed.), Academic Press, New York, 1972.

110. C. Nastasescu, Décomposition primaire dans les anneaux semi-artiniens, J. Algebra 14 (1970), 170-181.

111. C. Nastasescu and T. Albu, Décomposition primaire des modules, J. Algebra 23 (1972), 263-270.

112. C. Nastasescu and N. Popescu, Anneaux semi-artiniens, Bull. Soc. Math. France 96 (1968), 357-368.

113. C. Nastasescu and N. Popescu, On the localization ring of a ring, J. Algebra 15 (1970), 41-56.

114. Nguyen-Trong-Kham, D-anneaux. I, Rev. Roumaine Math. Pures Appl. 17 (1972), 1671-1680.

115. Nguyen-Trong-Kham, Théorie de la décomposition primaire dans les D-anneaux. II, Rev. Roumaine Math. Pures Appl. 18 (1973), 65-76.

116. Nguyen-Trong-Kham, D-modules, Rev. Roumaine Math. Pures Appl. 18 (1973), 283-289.

117. S. Page, Properties of quotient rings, Canad. J. Math. 24 (1972), 1122-1128.

118. Z. Papp, S-modules and torsion theories over artin rings, Arch. Math. 23 (1972), 598-602.

119. N. Popescu, "Categorii Abeliene", Editura Academiei R. S. R., Bucarest, 1971. (Roumanian)

120. N. Popescu, Le spectre à gauche d'un anneau, J. Algebra 18 (1971), 213-228.

121. N. Popescu, Les anneaux semi-noethériens, C. R. Acad. Sc. Paris 272 (1971), A1439-A1441.

122. N. Popescu, Théorie primaire de la décomposition dans les anneaux semi-noethériens, J. Algebra 23 (1972), 482-492.

123. N. Popescu and T. Spircu, Quelques observations sur les épimorphismes plats (à gauche) d'anneau, J. Algebra 16 (1970), 40-59.

124. M. Rayar, Small and cosmall modules, Ph. D. Thesis, Indiana Univeristy, 1971.

125. J. Raynaud, Sur la théorie de la localisation, thèse (3ème cycle) Université Claude-Bernard (Lyon I), 1971.

126. J. Raynaud, Localisations et anneaux semi-noethériens a droite, Publ. Dép. Math. Lyon 8-3, 77-112.

127. J. Raynaud, Localisations stables à droite et anneaux semi-noethériens à droite, C. R. Acad. Sc. Paris 275 (1972), A13-A16.

128. J. Rotman, A characterization of fields among integral domains, An. Acad. Brasil Ci. 32 (1960), 193-194.

129. R. Rubin, On exact localization, to appear, Pacific J. Math.

130. R. Rubin, Absolutely torsion-free rings, Bull. Amer. Math. Soc. 78 (1972), 854-856.

131. R. Rubin, Absolutely torsion-free rings, Pacific J. Math. 46 (1973), 503-514.

132. R. Rubin, Semi-simplicity relative to kernel functors, preprint 1973.

133. E. A. Rutter, Jr., Torsion theories over semiperfect rings, Proc. Amer. Math. Soc. 34 (1972), 389-395.

134. D. Salles, Anneaux semi-artiniens non commutatifs, C. R. Acad. Sc. Paris 275 (1972), A1223-A1225.

135. D. F. Sanderson, A generalization of divisibility and injectivity in modules, Canad. Math. Bull. 8 (1965), 505-513.

136. F. L. Sandomierski, Semisimple maximal quotient rings, Trans. Amer. Math. Soc. 128 (1967), 112-120.

137. W. Schelter, Rings of quotients, Ph. D. thesis, McGill University, 1972.

138. W. Schelter and P. Roberts, Flat modules and torsion theories, Math. Z. 129 (1972), 331-334.

139. T. Shores, The structure of Loewy modules, J. Reine Angew. Math. 254 (1972), 204-220.

140. T. Shores, Decomposition of torsion classes, preprint 1972.

141. T. Shores, Loewy series of modules, J. Reine Angew. Math. 265 (1974) 183-200.

142. L. Silver, Noncommutative localization and applications, J. Algebra 7 (1967), 44-76.

143. S. K. Sim, Some results on localizations of noncommutative rings, Ph. D. thesis, University of Western Ontario, 1973.

144. S. K. Sim, Prime torsion class with property (T), preprint 1973.

145. B. Stenström, Pure submodules, Ark. für Mat. 7 (1967), 159-171.

146. B. Stenström, On the completion of modules in an additive topology, J. Algebra 16 (1970), 523-540.

147. B. Stenström, "Rings and modules of quotients", Lecture Notes in Mathematics 237, Springer-Verlag, Berlin, 1971.

148. H. Storrer, Rings of quotients of perfect rings, Math. Z. 122 (1971), 151-165.

149. H. Storrer, On Goldman's primary decomposition, in "Lectures on rings and modules", Lecture Notes in Mathematics 246, Springer-Verlag, Berlin, 1972.

150. H. Storrer, Epimorphic extensions of noncommutative rings, Comm. Math. Helv. 48 (1973), 72-86.

151. H. Tachikawa, On splitting of module categories, Math. Z. 111 (1969), 145-150.

152. M. Teply, Torsion free injective modules, Pacific J. Math. 28 (1969), 441-453.

153. M. Teply, Some aspects of Goldie's torsion theory, Pacific J. Math. 29 (1969), 447-459.

154. M. Teply, Homological dimension and splitting torsion theories, Pacific J. Math. 34 (1970), 193-205.

155. M. Teply, Direct decomposition of modules, Math. Japon. 15 (1970), 85-90.

156. M. Teply, Torsion free projective modules, Proc. Amer. Math. Soc. 27 (1971), 29-34.

157. M. Teply, On noncommutative splitting rings, J. London Math. Soc. (2) 4 (1971), 157-164; also corrigendum, J. London Math. Soc. (2) 6 (1973), 267-268.

158. M. Teply, A class of divisible modules, Pacific J. Math. 45 (1973), 653-688.

159. D. Turnidge, Torsion theories and semihereditary rings, Proc. Amer. Math. Soc. 24 (1970), 137-143.

160. D. Turnidge, Torsion theories and rings of quotients of Morita equivalent rings, Pacific J. Math. 37 (1971), 225-234.

161. Y. Utumi, On quotient rings, Osaka J. Math. 8 (1956), 1-18.

162. A. Van der Water, A property of torsion-free modules over left Ore domains, Proc. Amer. Math. Soc. 25 (1970), 199-201.

163. C. Vinsonhaler, A note on two generalizations of QF-3, Pacific J. Math. 40 (1972), 229-233.

164. C. L. Walker and E. A. Walker, Quotient categories and rings of quotients, Rocky Mtn. J. Math. 2 (1972), 513-555.

165. C. N. Winton, On the derived quotient module, Trans. Amer. Math. Soc. 154 (1971), 315-321.

166. O. Zariski and P. Samuel, "Commutative Algebra", Van Nostrand, New York, 1958.

167. J. Zelmanowitz, Semiprime modules with maximum conditions, J. Algebra 25 (1973), 554-574.